Monographs in Electrical and Electronic Engineering

Editors: P. Hammond and D. Walsh

WALTER SCHOTTKY

(1886 - 1976)

who made significant contributions in the fields of electron physics,
radio engineering, statistical thermodynamics, and solid-state physics.

Metal-semiconductor contacts

E. H. RHODERICK

CLARENDON PRESS · OXFORD
1980

Oxford University Press, Walton Street, Oxford OX2 6DP

OXFORD LONDON GLASGOW
NEW YORK TORONTO MELBOURNE WELLINGTON
NAIROBI DAR ES SALAAM CAPE TOWN
KUALA LUMPUR SINGAPORE JAKARTA HONG KONG TOKYO
DELHI BOMBAY CALCUTTA MADRAS KARACHI

© *Oxford University Press 1978*

ISBN 0 19 859365 1

first published 1978

new as paperback edition 1980

PRINTED BY
THOMSON LITHO
EAST KILBRIDE, SCOTLAND

Preface

Twenty years have elapsed since the appearance of Henisch's classic book *Rectifying semiconductor contacts*. The intervening period has been one of intense activity in the field of semiconductor contacts, partly because of their immense importance in semiconductor technology and partly due to the development of new experimental techniques, so that we now have a much fuller understanding of the basic contact phenomena than we did twenty years ago. Although much of Henisch's book remains relevant today, it seems an appropriate moment for the publication of a book which embodies the developments of the past two decades.

I have not attempted to deal with every aspect of the subject in minute detail; rather, my aim has been to cover the basic principles fairly fully and to leave the reader who wishes to explore some particular facet in greater detail to do so with the help of the references provided. I have also deliberately not dealt with the practical applications of contacts beyond discussing methods of fabricating them, though I have tried to bring out the practical implications of the basic phenomena wherever possible. Applications may change, but the fundamental principles remain. The reader meeting the subject for the first time can obtain an overall view from Chapter 1, which constitutes a synopsis of the whole book and which contains cross-references to enable him to pursue any particular aspect more fully.

I must thank all those members of the Solid-State Electronics Group at UMIST who have contributed to the writing of this book, either directly through discussions or indirectly through their experimental work, and especially Dr. D.C. Northrop who read the entire manuscript and suggested many improvements. Dr. J.C. Inkson kindly read the section on intimate contacts and Dr. B. Hamilton the section dealing with deep traps. I must also thank Dr. V.L. Rideout of IBM for sending me a copy of a bibliography on metal—semiconductor contacts which he has prepared with A.H. Agajanian and which considerably eased the task of writing this book.

Finally, I owe a great debt of gratitude to Miss I.R. Winterburn, who typed the manuscript, for her uncanny instinct in unravelling my much-amended handwriting, and above all to my wife for her cheerful forbearance in putting up with hours of neglect.

Department of Electrical Engineering and Electronics   E.H.R.
University of Manchester Institute of Science and Technology
April 1977

Contents

List of Symbols

a	$= (q^2 N_d / 2\varepsilon_s kT)^{\frac{1}{2}}$ (eqn (3.7))
A	Richardson constant corresponding to free-electron mass
A^*	Richardson constant corresponding to effective mass in semiconductor (eqn (3.11))
A^{**}	value of A^* corrected for quantum-mechanical reflection and phonon backscattering
b	temperature coefficient of φ_{b0}
C	differential capacitance per unit area ($= dQ/dV$)
D_e	diffusion constant for electrons
D_h	diffusion constant for holes
D_s	density of surface states ($eV^{-1}\ m^{-2}$)
e_e	probability per unit time of a trap emitting an electron
e_h	probability per unit time of a trap emitting a hole
E	electron energy
E_c	energy of bottom of conduction band in semiconductor
E_F	Fermi energy
E_F^m	Fermi level in metal
E_F^s	Fermi level in semiconductor
E_g	energy gap of semiconductor
E_i	intrinsic Fermi level of semiconductor
E_t	energy of trap
E_v	energy of top of valence band
E_0	defined by eqn (3.27a)
E_{00}	defined by eqn (3.23)
E_1	$= E_i + (kT/2q)\ \ln\ (\sigma_h/\sigma_e)$ (§ 4.4.1)
$\mathscr{E}$	electric-field strength in semiconductor
$\mathscr{E}_i$	electric-field strength in interfacial layer

$\mathcal{E}_{max}$ maximum electric-field strength in Schottky barrier

f probability of occupation of trap

$F(x)$ Dawson's integral $\left(= \exp(-x^2) \int_0^x \exp(y^2)\,dy \right)$

h Planck's constant

$\hbar$ Planck's constant divided by 2π

i $= \sqrt{-1}$

I current

J current density

J_0 reverse saturation current density

J_e current density due to electrons

J_h current density due to holes

J_{te} current density due to thermionic emission

J_r current density due to recombination in depletion region

J_g current density due to generation in depletion region

k Boltzmann's constant

k wave vector of electron (§ 3.3.1)

l mean free path of electrons

L thickness of quasi-neutral region

L_e diffusion length for electrons

m free-electron mass

m^* effective mass of electrons in semiconductor

n density of electrons in conduction band of semiconductor *or* ideality factor (eqn (3.14))

n_i intrinsic electron concentration

N_c effective density of states in conduction band of semiconductor

N_d donor density

N_t trap density

N_v effective density of states in valence band of semiconductor

p density of holes in valence band of semiconductor

p_0 equilibrium density of holes at edge of depletion region

q magnitude of electronic charge

Q_d charge per unit area due to uncompensated donors

Q_h charge per unit area due to holes

Q_m charge per unit area on surface of metal

Q_{ss} charge per unit area due to surface states

r_0 Thomas–Fermi screening distance

S area of contact

T absolute temperature

u $= q\psi/kT$ (Appendix B)

u_s value of u at surface

$\bar{v}$ mean thermal velocity of electrons or holes

v_d diffusion velocity (eqn (3.18))

v_r recombination velocity (§ 3.2.5)

V applied bias (positive for forward bias)

V_d diffusion voltage or 'band bending'

V_{d0} diffusion voltage at zero bias

V_i drop in potential across interfacial layer

V_r reverse bias ($= -V$)

w width of depletion region

x_m position of maximum of barrier

y length defined in Fig. 4.10 ($= w-\lambda$)

α $= \delta\varepsilon_s/(\varepsilon_i + q\delta D_s)$ (eqn (2.10))

α' empirical quantity denoting dependence of φ_b on $\mathcal{E}_{max}$ (§ 3.6.1)

β $= (\partial\varphi_e/\partial V)$ (eqn (3.12))

γ $= \varepsilon_i/(\varepsilon_i + q\delta D_s)$ (eqn (2.6))

γ_h hole-injection ratio

δ thickness of interfacial layer

ε_i permittivity of interfacial layer ($= \varepsilon_{ir}\varepsilon_0$)

ε_s permittivity of semiconductor ($= \varepsilon_{sr}\varepsilon_0$)

ε_s' effective permittivity of semiconductor for image force

ε_0 permittivity of free space

ζ_e quasi-Fermi level (or imref) for electrons

ζ_h quasi-Fermi level (or imref) for holes

ϑ electronegativity (§ 2.9.2)

λ length defined in Fig. 4.10

μ mobility of electrons

ν frequency of light wave

ξ $= E_c - E_F^s$

ρ surface-dipole moment per unit area

σ_e capture cross-section for electrons

σ_h capture cross-section for holes

τ_{ce} mean time between collisions for electrons

τ_{ch} mean time between collisions for holes

τ_{re} recombination time for electrons

τ_{rh} recombination time for holes

φ_b height of Schottky barrier (measured from E_F^m)

φ_{b0} barrier height for zero bias

φ_b^0 barrier height for zero electric field (flat-band barrier height)

φ_{bn}^0 flat-band barrier height for n-type semiconductor

φ_{bp}^0 flat-band barrier height for p-type semiconductor

φ_e effective barrier height ($= \varphi_b - \Delta\varphi_{bi}$)

φ_{e0} effective barrier height for zero bias

φ_m work function of metal

φ_m' work function of metal relative to conduction band of SiO_2 (§ 2.7.1)

φ_0 neutral level for surface states (§ 2.1.3)

$\Delta\varphi_{bi}$ lowering of barrier due to image force

$(\Delta\varphi_{bi})_0$ image-force lowering for zero bias

χ_s electron affinity of semiconductor

χ_s' electron affinity relative to conduction band of SiO_2 (§ 2.7.1)

ψ electrostatic potential

ω_b angular frequency of bias modulation

ω_s angular frequency of test signal

Note: MKS units are used throughout. All energies are measured in eV, so the magnitude of the potential energy of an electron associated with its electrostatic potential is equal to the potential measured in volts.

The density of surface states is sometimes measured in $J^{-1}m^{-2}$. If D_s is expressed in $eV^{-1}m^{-2}$ and D_s' in $J^{-1}m^{-2}$, then $D_s = qD_s'$.

'I observed that in several natural and
artificial sulphides, both single crystals and
rough samples, the resistance was different by
up to 30% depending on the direction and intensity of the current.... The difficulty in these
experiments lay in obtaining reliable contacts.
I used mercury contacts.'

F. Braun (1874), *Poggendorff's Annalen*, **153**, 556

'You pinched two lead soldiers from your small
brother, melted them down over the fire, tossed in
a handful of flowers of sulphur, and produced what
was called a crystal.... You fiddled about, touch-
ing the crystal with a bit of thin wire called a
cat's whisker, and suddenly there in your head-
phones was the voice of a man....'

From a gramophone record issued by the British
Broadcasting Corporation in 1972 to commemorate
50 years of broadcasting.

1. Introduction and synopsis *

Our knowledge of metal—semiconductor contacts goes back more than a hundred years to the early work of Braun (1874) who discovered the asymmetric nature of electrical conduction between metal contacts and semiconductors such as copper and iron sulphide. Although the rectification mechanism was not understood, contacts between metal points and metallic sulphides were used extensively as detectors in early experiments on radio, and it seems likely that Lodge's 'coherer' (1890) must have relied for its action on the conduction properties of metal particles separated by oxide films. In 1906 Pickard took out a patent for a point-contact detector using silicon, and in 1907 Pierce published rectification characteristics of diodes made by sputtering metals onto a variety of semiconductors. The rapid growth of broadcasting in the 1920s owed much to the 'cat's-whisker' rectifier which consisted of a tungsten point in contact with a crystal, usually of lead sulphide. The first copper-oxide plate rectifiers appeared at about the same time (Grondahl 1926, 1933).

The first step towards understanding the rectifying action of metal—semiconductor contacts was taken in 1931, when Schottky, Störmer, and Waibel showed that if a current flows the potential drop occurs almost entirely at the contact, thereby implying the existence of some sort of potential barrier. By this time quantum mechanics was firmly established, and in 1932 Wilson and others tried to explain the rectifying action in terms of quantum-mechanical tunnelling of electrons through a barrier, but it was soon realized that this mechanism predicted the wrong direction of easy current flow. In 1938 Schottky and, independently, Mott, pointed out that the observed direction of rectification could be explained by supposing that electrons passed over a barrier through the normal processes of drift and diffusion. According to Mott (1938), the potential barrier arose because of a difference between the work functions of the metal and semiconductor; he supposed that the barrier region was devoid of charged impurities so that the electric field was constant and the electrostatic potential varied linearly as the metal was approached. In contrast,

---

*This Chapter is intended to provide an overall view for those meeting the subject for the first time.

Schottky (1939) supposed that the barrier region contained a constant density of charged impurities so that the electric field increased linearly
and the electrostatic potential quadratically, in accordance with Poisson's
equation, as the metal was approached. Similar ideas about the role of the
space charge in determining the shape of the barrier were advanced by
Davydov (1939, 1941) in the USSR.

   A significant advance in our understanding of metal—semiconductor contacts came during the Second World War as a result of the use of silicon
and germanium point-contact rectifiers in microwave radar. This advance was
considerably helped by developments in semiconductor physics. Perhaps the
most important contribution during this period was Bethe's thermionic-
emission theory (1942), according to which the current is determined by the
process of emission of electrons into the metal, rather than by drift and
diffusion in the semiconductor as was supposed by Mott and Schottky.

   After the war, work on metal—semiconductor contacts was stimulated by
the intense activity in the field of semiconductor physics which led up to
the invention of the point-contact transistor, and attention was mainly
focused on point contacts as injectors of minority carriers. The demise of
the point-contact transistor switched attention towards extended-area contacts. The realization that evaporation of metal films in a high-vacuum
system could produce contacts which were much more stable and reproducible
than point contacts triggered off a great flurry of activity in the 1950s
and 1960s and laid the foundation for our present extensive knowledge of
the subject. This activity was inspired to a considerable extent by the
great importance of contacts in semiconductor technology and is largely
associated with the theoretical work of Bardeen, Crowell, and Sze and the
experimental work of Goodman, Archer and Atalla, Kahng, Mead, and
Cowley.

1.2. THE SCHOTTKY BARRIER

Schottky's assumption that the shape of the potential barrier is determined by a uniform space charge due to ionized impurities conforms fairly
closely to what usually occurs in practice, so that metal—semiconductor
contacts are often somewhat indiscriminately referred to as Shottky barriers, and contacts which have intentional rectifying properties as Shottky
diodes. Such barriers are parabolic in shape, as shown in Fig. 1.1 for an
n-type semiconductor. In the n-type case, the impurities are shallow donors
which are normally completely ionized at room temperature. The upward
bending of the bands due to the positive space charge gives rise to a

region which is depleted of conduction electrons as on the n-type side of a p—n junction. This region is variously referred to as the barrier region, the space-charge region, or the depletion region. In a p-type semiconductor, the bands bend downwards because of the negative charge on the ionized acceptors. This downward bending constitutes a barrier to holes, because in some ways they behave like bubbles and have difficulty in passing *underneath* a barrier.

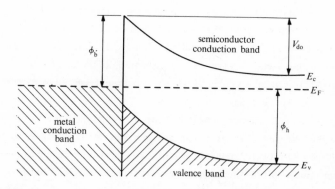

FIG. 1.1. Schottky barrier between a metal and an n-type semiconductor (zero bias).

The barrier arises because of the contact potential (difference in work functions) between the metal and the semiconductor. According to Mott (1938), the 'diffusion potential' or 'band bending' V_{d0} should be equal to the difference between the work functions ($\varphi_m - \varphi_s$), and the barrier height φ_b seen from the metal should consequently be given by the difference between the metal work function φ_m and the electron affinity of the semiconductor χ_s (see § 2.2.1). Measurements of barrier heights showed that φ_b is often almost independent of the choice of metal, and Bardeen (1947) explained this in terms of the existence of surface states on the semiconductor. These surface states effectively screen the interior of the semiconductor from the metal and absorb the contact potential difference (§ 2.2.2). If the density of surface states is very large, the barrier height should be independent of the work function of the metal, but in the general case φ_b should be a linear function of φ_m with a slope less than unity (§ 2.3.1).

The Bardeen model assumes the existence of a thin insulating layer between the metal and the semiconductor. This corresponds closely to the

actual situation if the contact is made by depositing a thin metal film
onto the etched surface of a semiconductor, since the chemical etch invariably leaves a thin oxide layer about 10—20 Å thick on the surface. The
experimental data on barrier heights obtained by depositing different metals
on etched surfaces of a particular semiconductor show a very poor correlation with the work function of the metal. This is partly because the barrier heights depend on the precise method used to prepare the surface of
the semiconductor and so show a fair amount of scatter between one set of
measurements and another, and partly because of uncertainties in the work
functions of the metals. All that can be said with confidence is that
metals with high work functions tend to give high barriers and metals with
low work functions tend to give low barriers (§ 2.7).

It is possible to make contacts which are free from any interfacial
insulating layer by cleaving a semiconductor so as to expose a fresh surface in an ultra-high vacuum, and then rapidly depositing a metal film by
evaporation before there is time for the semiconductor surface to be contaminated. Such 'intimate' contacts are more difficult to analyse theoretically than those which incorporate an interfacial layer, because the
latter 'decouples' the semiconductor from the metal to such an extent that
the surface states are not influenced very much by the metal. With intimate contacts, however, the surface states are strongly affected by the
metal, and the metal and semiconductor have to be regarded together as a
single quantum-mechanical system. The computational problems are formidable, and there is little available in the way of quantitative theoretical
prediction. The experimental situation is, however, more satisfactory than
that with 'real' contacts because of the improved reproducibility of the
measurements (§ 2.8); the data show that the barrier height is not a monotonic function of the metal work function, and it is clear that the
detailed band structure of the metal plays a role (§ 2.9.2). Most of the
data on cleaved surfaces refer to silicon and show much less dependence on
the choice of metal than with etched surfaces; the majority of metals give
barrier heights between 0.7 and 0.9 eV. If an explanation in terms of surface states is attempted, it is evident that the density of surface states
on cleaved surfaces is much higher than on etched surfaces (§ 2.9.3).

1.3. THE CURRENT/VOLTAGE CHARACTERISTIC

The most important process determining the current in a metal—semiconductor
contact to which a bias voltage is applied is the flow of electrons over
the top of the barrier from the semiconductor to the metal and vice versa.

In the absence of bias the current density J_{sm} due to electrons passing from the semiconductor to the metal must be equal and opposite to the current density J_{ms} due to electrons passing from the metal to the semiconductor. This is shown in Fig. 1.2(a) in which the arrows represent the direction of electron flow and not the conventional electric current. If a bias voltage is applied to the contact so that the metal is positive (assuming an n-type semiconductor), the bands in the semiconductor are raised in energy relative to those in the metal and the electric field in the barrier decreases. Because the resistivity of the metal is many orders of magnitude lower than that of the depletion layer, the decrease in electric field takes place entirely within the semiconductor barrier region and the shape of the barrier changes as shown in Fig. 1.2(b). The diffusion potential V_d is decreased compared with the zero-bias case, and the number of electrons able to surmount the barrier from the semiconductor increases so that J_{sm} increases. However, the barrier height φ_b seen from the metal does not change and J_{ms} remains constant so that there is a net flow of electrons into the metal. Because there is a relatively copious supply of electrons in the interior of the semiconductor, J_{sm} can increase

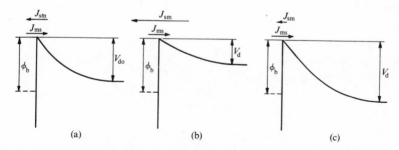

(a)   (b)   (c)

FIG. 1.2. Schottky barrier (a) under zero bias, (b) under forward bias, and (c) under reverse bias, showing the two components of the current.

by many orders of magnitude as the applied voltage increases. This is the 'easy' or 'forward' direction of current flow. If the bias voltage is reversed so that the semiconductor is positive, the energy bands in the semiconductor are lowered relative to those in the metal and the diffusion potential V_d is increased (Fig. 1.2(c)) so the number of electrons able to surmount the barrier from the semiconductor decreases. However, the barrier height φ_b seen from the metal does not change so J_{ms} remains constant and there is a net flow of electrons into the semiconductor. However large the applied voltage may be, J_{sm} cannot be less than zero, so the net current

saturates at the value J_{ms} which is independent of bias voltage. This is
the 'hard' or 'reverse' direction of current flow.

For an electron from the interior of the semiconductor to enter the
metal it must first pass through the depletion region. In doing so its
motion is governed by the usual processes of diffusion and drift in the
(opposing) electric field. When it reaches the metal, it suddenly receives
a large amount of momentum (corresponding to the Fermi velocity in the
metal) normal to the interface and passes into a narrow cone whose axis is
perpendicular to the boundary (Fig. 1.3). This latter process is governed
by the number of Bloch states in the metal which can communicate
with the semiconductor. These two processes are effectively in series,
and the current is determined by whichever presents the greater impediment
to current flow (§ 3.2.1). At first it was assumed by Mott and Schottky
that the drift and diffusion of electrons in the barrier region constitute
the bottleneck, and this is the basis of the 'diffusion theory' (§ 3.2.2).
Subsequently Bethe proposed that it is the actual emission of electrons
from the semiconductor into the metal which limits the current and, because
the process resembles thermionic emission, Bethe's theory is referred to as
the 'thermionic-emission theory' (§ 3.2.3).

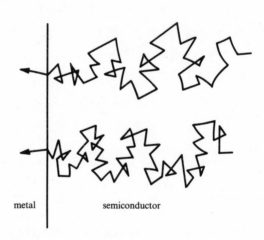

metal   semiconductor

FIG. 1.3. Typical trajectories of electrons passing from the semiconductor
to the metal showing the effects of drift and diffusion in the semiconductor followed by thermionic emission into a narrow cone in the metal.

It appears that the thermionic-emission theory adequately describes the
conduction mechanism in contacts to high-mobility semiconductors such as

silicon or gallium arsenide (§ 3.2.8). The current/voltage relationship for
this case can be shown to be given by

$$J = A^{**}T^2 \exp(-q\varphi_b/kT) \{\exp(qV/kT) - 1\}. \qquad (1.1)$$

Here J is the current density per unit area, and A^{**} is the Richardson constant modified to take account of the effective mass in the semiconductor
and various other corrections discussed in § 3.2.7. A^{**} is usually less
than the normal value of 12×10^5 A m^{-2} K^{-2}.

For bias voltages in excess of $3kT/q$ the last term in the bracket is
negligible, and the current density should be proportional to $\exp(qV/kT)$.
This ideal behaviour is never observed in practice, but instead the current
is usually found to vary as $\exp(qV/nkT)$, where n is approximately constant
and is greater than unity; n is often referred to as the 'ideality factor'.
In a good rectifier n may be as low as about 1.02. There are several possible reasons why n may exceed unity. In practice, the barrier height φ_b
usually changes slightly with the application of bias. This may be due to
the effect of the image force (§ 2.4) or to the existence of an interfacial
layer (§ 2.3.4). Both of these have the effect of increasing φ_b when a forward bias is applied so that the current increases less rapidly with V;
this is equivalent to a value of n greater than unity. Another possible
cause of deviation from the ideal current/voltage relationship may be the
recombination of electrons and holes in the depletion region, which is
often important with high barriers and in materials with short lifetimes
such as gallium arsenide. This gives rise to a component of current with
$n = 2$, over and above the thermionic-emission current, which can be very
pronounced at low values of applied bias (§ 3.4). At high values of applied
bias, corresponding to large current densities, there may be an appreciable
rise in n because the effect of drift and diffusion in the barrier region
is no longer negligible and the pure thermionic-emission theory does not
apply (§ 3.2.8).

A further complication of the current/voltage relationship in practical
diodes is that, for large forward currents, the voltage drop across the
series resistance R_s associated with the neutral region of the semiconductor (between the depletion region and the ohmic contact) causes the actual
voltage developed across the barrier region to be less than the voltage
applied to the terminals of the diode. Instead of obeying the ideal equation (1.1), the current density is proportional to $\exp\{q(V-IR_s)/kT\} - 1$,
where I is the current through the diode, and a plot of $\ln I$ against V

deviates from a straight line at high forward voltages as shown in Fig.
1.4. The series resistance can be found experimentally in two ways. The
most direct method is to apply a sufficiently large forward bias (greater
than φ_b) for the bands in the semiconductor to become flat. The depletion
region will disappear, and the resistance of the barrier region will now
be negligible compared with R_s; a measurement of I and V then allows R_s to
be obtained directly. The other method makes use of the plot of ln I against
V shown in Fig. 1.4. At large forward voltages, the horizontal displacement
ΔV (for a given I) between the actual curve and the extrapolation of the
linear region gives the voltage drop IR_s across the neutral region and, by
plotting ΔV against I, the value of R_s may be determined.

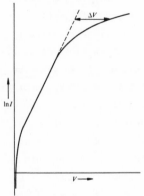

FIG. 1.4. Practical current/voltage characteristic of a rectifier showing
the effect of series resistance.

If the donor density is increased (assuming an n-type semiconductor),
the curvature of the bands increases in accordance with Poisson's equation
and the barrier region becomes thinner. A stage is eventually reached
(corresponding to a donor density rather greater than about 5×10^{23} m^{-3}
in silicon at room temperature) at which electrons can penetrate the tip
of the barrier by quantum-mechanical tunnelling. The effect is as if the
barrier height were reduced, and the current for a given value of bias is
increased. If the donor density is further increased, electrons can tunnel
through the barrier even though their energy may be substantially below
the top of the barrier; the current is still an exponential function of
the forward voltage but with an n value considerably greater than unity.
This is known as thermionic-field emission. The energy at which most of the
electrons emerge through the barrier is determined by a competition between the decrease in the number of electrons and the increase in their

tunnelling probability as the electron energy increases; it therefore depends on the temperature, as do the current density and the n value. Finally, at low temperatures, if the donor density is so large that the semiconductor becomes degenerate and the Fermi level lies within the conduction band (e.g. for donor densities in excess of 10^{25} m^{-3} in silicon), electrons may actually tunnel at the Fermi level. This is known as field emission (§ 3.3.1 and Fig. 3.4).

The straightforward thermionic-emission theory predicts that, under reverse bias, the current density should saturate at a value $J = A^{**} T^2$ $\exp(-q\varphi_b/kT)$. This is never found to happen in practice; instead, the reverse current usually shows a gradual increase with increasing reverse bias which eventually becomes a very rapid increase generally described as breakdown. The gradual increase is usually due to a bias dependence of the barrier height which may arise from image-force lowering, from the penetration of the metal–electron wave functions (§ 3.6.1), or from the presence of an interfacial layer (§ 3.8). The effect of reverse bias is always to decrease the barrier height and therefore to increase the current. Alternatively, the gradual increase may be due to the thermal generation of hole–electron pairs in the depletion region, which is often important in large-band-gap low-lifetime semiconductors. The current due to this mechanism increases with bias voltage because the width of the depletion region increases as $|V|^{\frac{1}{2}}$ (§ 3.6.3). In moderately doped semiconductors, a gradual increase in reverse current with applied bias may also result from the onset of thermionic-field emission, which becomes significant at lower doping levels in the reverse direction than in the forward direction because the application of a reverse bias of a few volts can cause the barrier to become thin enough for tunnelling to take place (§ 3.6.2).

At higher doping concentrations (e.g. greater than about 5×10^{23} m^{-3} in silicon at room temperature), thermionic-field emission may become firmly established and can give rise to a rapid increase of current with reverse bias which is analogous to Zener breakdown in p–n junctions. Even at only moderately high doping levels (say about 10^{23} m^{-3} in silicon), thermionic-field emission may cause reverse breakdown at the edge of the metal contact because the electric field is liable to be very high (and the barrier much thinner) owing to crowding of the field lines. This premature breakdown may be avoided by the use of a guard-ring of diffused p-type material (in the case of an n-type semiconductor) underneath the periphery of the contact. The presence of the p-type material eliminates the high field at the edge and places a reverse-biased p–n junction in parallel with the

Schottky barrier (§ 3.6.2). Finally, for very high donor concentrations (in excess of 10^{24} m^{-3} for silicon) and at low temperatures, field emission may take place. If this happens, the 'reverse' current may actually exceed the 'forward' current, so that the direction of rectification is inverted as originally predicted by Wilson (§ 3.6.2).

Diodes made from moderately doped semiconductors with a donor density of less than about 10^{23} m^{-3} (for silicon) and a guard-ring to avoid edge effects will withstand comparatively high reverse-bias voltages and will eventually show avalanche breakdown as in a p–n junction. Breakdown voltages as high as 1 kV may be obtained in silicon Schottky diodes with a donor density of 10^{20} m^{-3}.

1.4. COMPARISON WITH p–n JUNCTIONS

An important feature of conduction in a forward-biased Schottky barrier is that the current is carried almost entirely by electrons (assuming an n-type semiconductor), even if the semiconductor is only lightly doped. This contrasts with the case of a p–n junction, where the current is carried predominantly by those carriers which emanate from the more heavily doped side. A Schottky diode on n-type material therefore corresponds to a p–n junction in which the p side is the more lightly doped side; i.e. to a p$^-$–n junction. When such a p$^-$–n junction is forward biased, electrons are injected into the p side and, if the bias is suddenly reversed, the injected electrons must be removed before the junction assumes a high-resistance state, so that an appreciable current flows for a short time in the reverse direction. This phenomenon is known as minority-carrier storage. The analogue of this process in a Schottky diode is the injection of electrons into the metal. These electrons can, in principle, be drawn back into the semiconductor if the bias is suddenly reversed, but only if they retain sufficient energy to surmount the barrier. When electrons are injected into the metal under forward bias, they are 'hot' electrons in the sense that their energy exceeds the Fermi energy in the metal by the barrier height (§ 3.2.1). This surplus energy is rapidly lost by collision with other electrons in the metal within about 10^{-14} s, so they can only return to the semiconductor within a time of this order after the reversal of the bias. The usual minority-carrier-storage-effect characteristic of p–n junctions is therefore virtually absent in Schottky diodes, and the recovery time is determined by other considerations (§ 3.7).

In a p$^-$–n junction a small fraction of the current is carried by holes. This process can also take place in a Schottky diode, but the fraction is exceedingly small (typically of the order of 10^{-4} of the electron current). This is partly because the barrier to electron flow φ_b is usually smaller than the barrier to hole flow φ_h (see Fig. 1.1), and partly because the thermionic-emission process is intrinsically more efficient than the diffusion process which determines the hole current (§ 3.5.1). This process, known as hole injection, can also give rise to a current transient when the bias is suddenly reversed but, although the lifetime of the holes is many orders greater than the time taken by the 'hot' electrons to lose their excess energy, the injection ratio J_h/J_e is so small that even this contribution to the recovery time is still negligible compared with the recovery time of a p–n junction. In practice the recovery time of Schottky diodes is determined by their RC time constant rather than by electronic processes and, for this reason, they are used extensively as fast-switching diodes and as microwave mixers (§ 3.7 and Appendix C).

In a p–n junction the injection of minority carriers into the more lightly doped side brings about an increase in the majority-carrier concentration to maintain charge neutrality. This process is known as conductivity modulation and plays an important role in reducing the series resistance of high-voltage rectifiers. Conductivity modulation can in principle occur in Schottky diodes as a consequence of hole injection, but it is generally negligible because of the small injection ratio. It can, however, be observed in diodes made from high-resistivity semiconductors and metals that give large barrier heights (§ 3.5.1).

As we have already mentioned, the thermionic-emission process is intrinsically more efficient than the diffusion process (Appendix C) so that, for a given diffusion potential V_{do}, the saturation current of a Schottky diode is several orders of magnitude greater than that of a p–n junction. Furthermore, one can obtain much smaller diffusion potentials in Schottky diodes than in p–n junctions made from the same material, and these two factors taken together make the saturation-current density of a Schottky diode exceed that of a p–n junction by a very large factor, as much as seven orders of magnitude or more. This can be expressed differently by saying that for the same forward-current density the bias applied to a p–n junction must exceed that applied to a Schottky diode by as much as 0.5 V (Appendix C), so that Schottky diodes are particularly suitable for use as low-voltage high-current rectifiers.

1.5. CAPACITIVE EFFECTS

Like a p–n junction, a Schottky barrier exhibits capacitance. With one

exception, the capacitance of a Schottky barrier in an n-type semiconductor is virtually identical to that of an abrupt p^+–n junction, since the
p^+ side of the latter behaves in many respects like a metal. The exception
concerns the diffusion capacitance. If an alternating voltage (superimposed
on a d.c. basis) is applied to a p^+–n junction, the injection of holes into
the n side during the forward halfcycle is followed by extraction during
the negative halfcycle, provided the d.c. bias is such that the a.c. component swings the total voltage into forward bias; this contributes a
capacitance known as the diffusion capacitance. The diffusion capacitance
is very closely linked with the phenomenon of minority-carrier storage. As
we have already seen, there is no minority-carrier storage in a Schottky
diode, so there is no diffusion capacitance either.

   The capacitance of a Schottky barrier is associated with its depletion
region, which in some respects resembles a parallel-plate capacitor with a
separation between the plates that increases when reverse bias is applied.
The capacitance is usually measured by superimposing a small alternating
voltage on a reverse d.c. bias, which yields the differential capacitance
$C(=dQ/dV)$. The differential capacitance is therefore not constant but
decreases as the reverse bias V_r is increased. It can be shown that C is
proportional to $(V_r+V_{d0})^{-\frac{1}{2}}$ as in the case of a p–n junction, so that a plot
of C^{-2} against V_r should be a straight line with an intercept of $-V_{d0}$ on the
voltage axis (§ 4.1.2). This forms the basis of a method of measuring barrier heights (§ 2.6.3). It can also be shown that for an n-type semiconduc-
tor the slope of the line is equal to $2/\varepsilon_s qN_d S^2$, where ε_s is the permittivity of the semiconductor and S the area of the contact. This was first
pointed out by Schottky (1942) and gives an easy method of measuring N_d.
If N_d is not constant, the graph of C^{-2} against V_r is not linear, but the
slope at any point is still given by $2/\varepsilon_s qN_d S^2$, where N_d is now the donor
density at the edge of the depletion region. The width w of the depletion
region can be obtained from $C = \varepsilon_s S/w$, so that N_d can be found as a function of w. This forms the basis of a very convenient method of measuring
impurity distributions (§ 4.3). The bias dependence of a reverse-biased
Schottky barrier can also be exploited as a means of achieving a voltage-
controlled variable capacitance.

   Sometimes the depletion region of a Schottky barrier may contain deep
traps associated with crystal defects or with impurities other than the
shallow donors or acceptors with which the semiconductor is doped. These
traps can have time constants as long as many seconds or even tens of
minutes. Such traps can have a very complicated effect on the capacitance,

depending on the relationship between the reciprocals of their time constants and the frequency of the measuring signal and the rate of change of
the d.c. bias (§ 4.4.2). In favourable circumstances, measurements of the
capacitance as a function of reverse-bias voltage and of time can yield not
only the concentration of the traps but also their time constant and energy
relative to the band edges (§ 4.4.3). The occupation of the traps can be
changed by external stimuli such as light or changes of temperature
(§ 4.4.4), or by the application of a forward bias (§ 4.5.2). These techniques form the basis of a new and very fertile branch of deep-level
spectroscopy.

1.6. PRACTICAL CONTACTS

While plate rectifiers were made by the oxidation of copper, the early
radio-frequency detectors almost exclusively involved point contacts. Point-
contact rectifiers are notoriously unstable and difficult to reproduce, and
the real advance in the understanding and application of metal—semiconductor
contacts came with the introduction of extended-area contacts incorporating
evaporated-metal films. The majority of contacts are now made in a conventional vacuum system by evaporating the metal onto a semiconductor surface
prepared by cutting, polishing, and etching, although it is becoming in-
creasingly common to use a very-high-vacuum system fitted with an ion pump.
Such contacts invariably contain a thin oxide film between the metal and
semiconductor and are therefore not ideal, but they are sufficiently near to
the ideal for most practical purposes. If contacts without an interfacial
film are required for research, they can be made by cleaving a crystal in
an ultra-high-vacuum system while the evaporation is in progress (§ 5.1.2).
Contacts can also be made by sputtering (§ 5.1.3) and by plating (§ 5.1.4).

In device applications contacts are always subjected to heat treatment,
sometimes deliberately to promote mechanical adhesion and sometimes un-
avoidably because other steps in the manufacture of the device involve
elevated temperatures. Even at comparatively low temperatures ($\sim$ 200°C),
metallurgical changes can take place so that the contact is far from being
an abrupt metal—semiconductor junction. In rectifying contacts such changes
usually result in I/V characteristics which are very far from ideal.

The nature of the metallurgical changes can be studied using the techniques of Rutherford backscattering, Auger electron spectroscopy, and
secondary-ion mass spectrometry (§ 5.2). It is usually difficult to relate
the degradation of the I/V characteristic to the observed metallurgical
changes. Those cases where correlation is possible usually involve either

the production of a heavily doped surface layer in the semiconductor due
to the migration of metal atoms or to the generation of charged defects,
so that tunnelling may take place, or the production of an insulating interfacial layer (§ 5.2).

In the case of silicon, an important class of contacts involves those
metals which form stoichiometric silicides. The majority of metals, including all the transition metals, fall into this category (§ 5.3). Nearly all
silicides exhibit metallic conductivity so that, if a silicide is formed as
a result of heat treatment, the silicide—silicon junction behaves like a
metal—semiconductor contact. As the interface is formed some distance below
the original surface of the silicon, it is free from contamination; contacts
made in this way generally show very stable electrical and mechanical
properties. The kinetics of silicide formation have been extensively studied
(§ 5.3.1). A surprising result is that the barrier height is virtually independent of the particular compound formed adjacent to the silicon
(§ 5.3.2).

It is often desirable to be able to control the barrier height. A limited degree of control is available through the choice of metal, but the
range of barrier heights obtained in this way is restricted by the requirement that the metal should have metallurgical properties compatible with
the thermal treatment to which it is subjected during processing. A number
of methods are available in principle, but the only practicable method
seems to be that of Shannon (§ 5.4) who produced a thin highly doped layer
near the surface of the semiconductor by ion bombardment. Doping with impurities of the same polarity as those in the bulk of the semiconductor
produces a thinner barrier through which carriers can tunnel by thermionic-
field emission; this is equivalent to a reduction in barrier height. Impurities of the opposite polarity, however, can produce a barrier with a
maximum occurring some distance from the metal and a height exceeding the
barrier associated with uniformly doped material.

1.7. OHMIC CONTACTS

Semiconducting devices and specimens used for the measurement of semiconductor parameters require ohmic contacts to which connections can be made.
The important feature of such contacts is that the voltage drop across
them must be negligible compared with the voltage drop across the device
or specimen, so that the contacts do not affect the I/V characteristic. In
principle, such contacts can be made by using a metal with a work function
less than the work function of an n-type semiconductor or greater than the

work function of a p-type semiconductor. However, there are very few metal—
semiconductor combinations which satisfy this condition. The vast majority
of ohmic contacts involve a thin layer of very heavily doped semiconductor
immediately adjacent to the metal, so that the depletion layer is so thin
that the carriers can readily tunnel through it (§§ 3.3.2 and 5.5). The
heavily doped layer may be formed separately from the deposition of the
metal contact or it may result from the deposition and subsequent heat
treatment of an alloy containing an element which acts as a donor or acceptor. Every laboratory tends to have its own favourite recipes.

1.8. THE PRESENT POSITION

The position at the time of writing (1977) is that we have a reasonably
good working picture of metal—semiconductor contacts that serves as a basis
for interpreting experimental results and also enables us to 'design' contacts to a limited extent to obtain particular electrical characteristics.
However, the finer details are still imperfectly understood. On the experimental side, the outstanding problem is that we still do not have a very
good understanding of the metallurgical nature of the metal—semiconductor
interface, which is generally far from being an abrupt change from an ideal
metal to an ideal semiconductor. There is a pressing need for more analytical techniques for studying interfaces, preferably with much better depth
resolution than those currently available.

On the theoretical side, we are only just beginning to appreciate that
barrier heights depend on the band structure of the metal and semiconductor
and not just on macroscopically measurable quantities such as work function
and electron affinity. The problem of what happens to surface states when
a semiconductor is overlaid by a metal is beginning to be tackled, but the
analytical problems are formidable. Progress on the theoretical front tends
to be largely confined to ideal metal—semiconductor interfaces, which as we
have seen do not bear much resemblance to reality. We have a long way to go
before we shall be able to combine a thorough understanding of the metallurgical problems with a satisfactory theoretical framework.

1.9. BIBLIOGRAPHY

The reader who wishes to pursue certain aspects of the subject more fully
than the treatment given in this book may find the following books and
review papers of value.

Books

The first book to appear on semiconductor contacts was Torrey and Whitmer's
Crystal rectifiers (1948), which describes in detail the wartime development of point-contact rectifiers for use as microwave detectors. It discusses the circuit aspects of microwave mixers and also throws some interesting
light on the state of semiconductor physics at the time. The classic book
on the subject is Henisch's *Rectifying semiconductor contacts* (1957), which
gives a comprehensive survey of the historical development of the subject.
It also discusses very fully the properties of point contacts as injectors
of minority carriers, which were of great importance for the development
of the point-contact transistor. Spenke's book *Electronic semiconductors*
(1958) contains a very full description of the diffusion and thermionic-
emission theories of rectification, and also discusses the physics of contact potentials in some detail. The books by Sze (1969) and Milnes and
Feucht (1972) also contain extended discussions of metal—semiconductor contacts, the latter containing a very useful list of references to papers on
ohmic contacts.

Review articles

Joffe (1945): A review (in English) of Schottky's papers.

Atalla (1966): A general review of the physics of Schottky barriers, with
some discussion of their applications.

Mead (1966): A brief review of Schottky barriers with a list of barrier
heights.

Rhoderick (1970): A general review of Schottky-barrier phenomena.

Padovani (1971): A review of current transport processes in contacts, with
special emphasis on tunnelling in gallium arsenide.

Rhoderick (1974): A general review of current-transport processes.

Grimmeiss (1974): A review of the use of Schottky barriers for deep-level
spectroscopy.

Sigurd (1974): A review of Rutherford backscattering techniques applied to
the study of metal—semiconductor interactions.

Conference proceedings

Ohmic contacts to semiconductors. (Montreal 1968). Published by the Electrochemical Society (1969). Contains a good deal of information about rectifying as well as ohmic contacts.

Metal—semiconductor contacts (Manchester 1974). Published by The Institute
of Physics (Conf. Series No. 22).

Symposium on Electrical and Structural Properties of Interfaces (1974).
Published in *J. Vac. Sci. Technol.* **11**, No. 6 (Nov.—Dec. 1974).

'Physics of Compound Semiconductor Interfaces' (San Diego 1976). Published in *J. Vac. Sci. Technol.* **13**, No. 4 (July—Aug. 1976).

2. The Schottky barrier

2.1. PRELIMINARIES: SOME SURFACE PROPERTIES OF SOLIDS

2.1.1. *The work function of a metal*

The work function φ_m of a metal is the amount of energy required to raise an electron from the Fermi level to a state of rest outside the surface of the metal (the so-called 'vacuum level'). If the work function is calculated using quantum mechanics (see, for example, Seitz 1940), it is found to consist of two parts — a volume contribution (which represents the energy of an electron due to the periodic potential of the crystal and the interaction of the electron with other electrons) and a surface contribution (due to the possible existence of a dipole layer at the surface). In general, the electron cloud around an atom at the surface is not symmetrically disposed in relation to the nucleus, so that the centres of positive and negative charge do not coincide. If the resulting dipole layer has an electric dipole moment ρ per unit area, there will be a difference of electrostatic potential of magnitude ρ/ε_0 between the vacuum and the interior of the metal. The change in energy $q\rho/\varepsilon_0$ of an electron due to this change in electrostatic potential constitutes part of the work function.

2.1.2. *The work function and electron affinity of a semiconductor*

As with a metal, the work function φ_s of a semiconductor is the difference in energy between the Fermi level and the energy of an electron at rest outside the surface. This is the quantity which determines the thermionic emission of electrons from a heated semiconductor. It may seem strange that the work function is defined in this way when there are usually no allowed energy levels within a semiconductor at the Fermi level, but it must be remembered that the work function is a statistical concept and represents a weighted average of the energies necessary to remove an electron from the valence or conduction band, respectively.

Another important surface parameters of a semiconductor is the electron affinity (χ_s). This is the difference in energy between an electron at the bottom of the conduction band and an electron at rest outside the surface. If the bands are flat (i.e. there is no electric field inside the semiconductor), the work function and electron affinity are related by

$$\varphi_s = \chi_s + \xi \tag{2.1}$$

where ξ is the energy difference between the Fermi level and the bottom of the conduction band. As in the case of a metal, the semiconductor electron-affinity has both a volume contribution and a surface-dipole contribution.

2.1.3. Surface states

Surface states arise because the surface of a crystal interrupts the perfect periodicity of the crystal lattice. Solutions of Schrödinger's equation exist which correspond to energy levels within the forbidden gap of the semiconductor and to imaginary values of the wave vector k; these wave functions are evanescent waves which decay exponentially with distance (they are in fact similar to the tunnelling wave functions discussed in § 3.3.1). They are therefore localized in space (unlike the Bloch waves which propagate throughout the crystal), and in a perfect semiconductor can only exist at a surface. These are the surface states predicted by Tamm and Shockley (see, for example, Many, Goldstein, and Grover 1965). A more detailed three-dimensional treatment of surface states (for example, Heine 1972) shows that they form a two-dimensional band with a continuous range of energies which may overlap the valence and conduction bands, though only those within the forbidden gap play a role in contact phenomena. Surface states of this sort are often described as 'intrinsic', which means that they exist at an ideally perfect surface.

The wave functions which make up the surface states are drawn from those which would constitute the valence and conduction bands of an infinite crystal, so that the densities of states in the valence and conduction bands are diminished close to the surface. It follows that the full complement of electrons necessary to make the surface as a whole electrically neutral can only be accommodated if the band of surface states is partly filled. This leads to the concept of a 'neutral level' φ_0, which is the level to which the surface states are filled when the surface is electrically neutral.[*] If states below φ_0 are empty, the surface has a net positive charge while, if states above φ_0 are filled, the surface has a net negative charge. The states below φ_0 are sometimes described as donor-like (positive when empty) and the levels above φ_0 as acceptor-like (negative when filled).

---

[*]The surface is said to be electrically neutral if the total volume taken up by the surface-state wave functions is neutral.

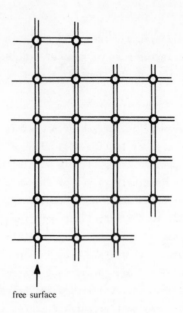

free surface

FIG. 2.1. Dangling bonds at the free surface of a covalent crystal. The atoms make double bonds with each other, but the atoms at the surface have no neighbours to bond to on one side.

It is sometimes helpful to look at surface states from the point of view of chemical bonds. Fig. 2.1 illustrates schematically a cubic semiconductor with a free surface, the atoms being bound to their neighbours by double covalent bonds. At the surface, the atoms have neighbours on one side only and, on the vacuum side, the valence electrons have no partners with which to form covalent bonds. Each surface atom therefore has associated with it an unpaired electron in a localized orbital which is directed away from the surface. Such an orbital is often spoken of as a 'dangling' bond; it can either give up its electron and act as a donor or accept another and act as an acceptor. According to this simple model, there should be twice as many surface states as there are surface atoms, and the neutrality condition should correspond to half of the surface states being occupied.

2.2. THE FORMATION OF A SCHOTTKY BARRIER

2.2.1. *The Schottky–Mott theory*

To see how a barrier may be formed when a metal comes into contact with a

semiconductor, suppose that the metal and semiconductor are both electrically neutral and separated from each other. The energy-band diagram is
shown in Fig. 2.2(a) for an n-type semiconductor with a work function less
than that of the metal; this is the most important case in practice, and
we suppose that there are no surface states present. If the metal and semiconductor are connected electrically by a wire, electrons pass from the
semiconductor into the metal and the two Fermi levels are forced into coincidence as shown in Fig. 2.2(b). The energies of electrons at rest outside the surfaces of the two solids are no longer the same, and there is an
electric field in the gap directed from right to left. There must be a
negative charge on the surface of the metal balanced by a positive charge
in the semiconductor. The charge on the surface of the metal consists
simply of extra conduction electrons contained within the Thomas—Fermi
screening distance (≈ 0.5 Å). Since the semiconductor is n-type, the positive charge will be provided by conduction electrons receding from the
surface leaving uncompensated positive donor ions in a region depleted of
electrons. Because the donor concentration is many orders of magnitude less
than the concentration of electrons in the metal, the uncompensated donors
occupy a layer of appreciable thickness w, comparable to the width of the
depletion region in a p—n junction, and the bands in the semiconductor are
bent upwards as shown in Fig. 2.2(b). The difference V_i between the electrostatic potentials outside the surfaces of the metal and semiconductor is
given by $V_i = \delta \mathscr{E}_i$, where δ is their separation and $\mathscr{E}_i$ the field in the gap.
If the metal and semiconductor approach each other, V_i must tend to zero
if $\mathscr{E}_i$ is to remain finite (Fig. 2.2(c)) and, when they finally touch (Fig.
2.2(d)), the barrier due to the vacuum disappears altogether and we are
left with an ideal metal—semiconductor contact. It is clear from the fact
that V_i tends to zero that the height of the barrier φ_b measured relative
to the Fermi level is given by

$$\varphi_b = \varphi_m - \chi_s. \tag{2.2}$$

In most practical metal—semiconductor contacts, the ideal situation shown
in Fig. 2.2(d) is never reached because there is usually a thin insulating
layer of oxide about 10–20 Å thick on the surface of the semiconductor.
Such an insulating film is often referred to as an interfacial layer. A
practical contact is therefore more like that shown in Fig. 2.2(c); however, the barrier presented to electrons by the oxide layer is so narrow
that electrons can tunnel through it quite easily, and Fig. 2.2(c) is

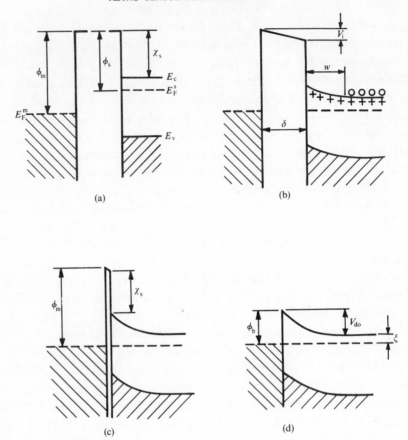

FIG. 2.2. Formation of a barrier between a metal and a semiconductor (a) neutral and isolated, (b) electrically connected, (c) separated by a narrow gap, (d) in perfect contact. O denotes electron in conduction band; + denotes donor ion.

almost indistinguishable from Fig. 2.2(d) as far as the conduction electrons are concerned. Moreover, the potential drop V_i in the oxide film is so small that eqn (2.2) is still a very good approximation.

Although it is usually attributed to Schottky, eqn (2.2) was first stated implicitly by Mott (1938) and will be referred to as the Mott limit. In obtaining it a number of important assumptions have been made, among them that the surface dipole contributions to φ_m and χ_s do not change when the metal and semiconductor are brought into contact (or, at least, that their difference does not change) and that there are no surface states.

The shape of the potential barrier depends on the charge distribution

within the depletion region. If the bottom of the conduction band is raised
by about 3 kT/q above its position in the bulk, the electron density is
reduced by an order of magnitude, and between this plane and the metal—
semiconductor interface the space charge is due entirely to the uncompensated donors. If we neglect the bending of the bands in the transition
region where the electron density is less than the donor concentration but
has not fallen by an order of magnitude (the so-called depletion approximation), the shape of the barrier will be determined entirely by the spatial
distribution of the donors.

   In the model first put forward by Schottky (1938) and elaborated by
Schottky and Spenke (1939), the semiconductor is assumed to be homogeneous
right up to the boundary with the metal, so that the uncompensated donors
give rise to a uniform space charge in the depletion region. The electric-
field strength therefore increases linearly with distance from the edge of
the depletion region in accordance with Gauss's theorem, and the electrostatic potential increases quadratically (see Fig. 2.3(a)). The resulting
parabolic barrier is known as a Schottky barrier.

   A somewhat different model was put forward by Mott (1938), who assumed
that the semiconductor had a thin layer devoid of donors immediately next
to the metal. The electric-field strength would be constant throughout
such a layer giving rise to an electrostatic potential which increases
quadratically at first and then linearly, as shown in Fig. 2.3(b). Such a
barrier is known as a Mott barrier. It is rarely encountered in practice.

   The foregoing description applies to an n-type semiconductor with work
function φ_s less than the work function φ_m of the metal. It will be seen
in Chapter 3 that such a contact behaves as a rectifier. If a similar argument is developed for the case when φ_s is greater than φ_m, one obtains a
band diagram of the form shown in Fig. 2.4(b). Clearly, if such a contact
is biased so that electrons flow from the semiconductor to the metal, they
encounter no barrier. If it is biased so that electrons flow in the reverse
direction, the comparatively high concentration of electrons in the region
where the semiconductor bands are bent downwards (usually referred to as an
accumulation region) behaves like a cathode which is easily capable of
providing a copious supply of electrons. The current is then determined by
the bulk resistance of the semiconductor. Such a contact behaves like an
ohmic contact.*

---

*By an 'ohmic' contact we mean one which has a sufficiently low resistance
for the current to be determined by the resistance of the bulk semiconductor (or I/V characteristic of a device) rather than by the properties of
the contact.

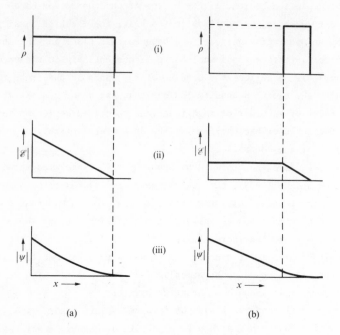

FIG. 2.3. (a) Schottky barrier, (b) Mott barrier (i) charge density (ii) electric-field strength (iii) electrostatic potential (ψ). For an n-type semiconductor, ψ is negative and the electron potential energy ($-q\psi$) positive.

In a p-type semiconductor for which φ_m exceeds φ_s, we obtain the band diagram shown in Fig. 2.4(c), which also represents an ohmic contact. The case of a p-type semiconductor for which φ_s exceeds φ_m is shown in Fig. 2.4(d). Bearing in mind that holes have difficulty in going underneath a barrier, one sees that Fig. 2.4(d) is the p-type analogue of Fig. 2.4(a) and gives rise to rectification.

Figs. 2.4(b) and (c) are very uncommon in practice, and the majority of metal—semiconductor combinations form rectifying or 'blocking' contacts. Unless the contrary is clearly stated, all subsequent discussions will centre round the case of n-type semiconductors with $\varphi_m > \varphi_s$, which is the most important case in practice.

2.2.2. *The effect of surface states*

Even if the assumption about the constancy of the surface dipole layer is incorrect, the barrier height φ_b should still depend on the metal work

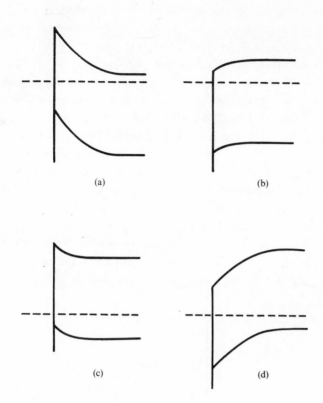

FIG. 2.4. Barriers for semiconductors of different types and work functions

n-type   (a) $\varphi_m > \varphi_s$ (rectifying)

          (b) $\varphi_m < \varphi_s$ (ohmic)

p-type   (c) $\varphi_m > \varphi_s$ (ohmic)

          (d) $\varphi_m < \varphi_s$ (rectifying)

function φ_m if the simple Mott theory (eqn (2.2)) is valid. But experimentally it is found that the barrier height is a less sensitive function of φ_m than eqn (2.2) would suggest and that, under certain circumstances, φ_b may be almost independent of the choice of metal.

An explanation of this weak dependence on φ_m was put forward by Bardeen (1947), who suggested that the discrepancy may be due to the effect of surface states. Suppose that the metal and semiconductor remain separated by a thin insulating layer as shown in Fig. 2.5 and that there is a continuous distribution of surface states present at the semiconductor surface characterized by a neutral level φ_0. In the absence of surface states,

the negative charge Q_m on the surface of the metal must be equal and opposite to the positive charge Q_d due to the uncompensated donors because the
junction as a whole is electrically neutral. In the presence of surface
states, the neutrality condition becomes $Q_m + Q_d + Q_{ss} = 0$, where Q_{ss} is
the charge in the surface states. The occupancy of the surface states is
determined by the Fermi level, which is constant throughout the barrier
region in the absence of applied bias, and for most purposes it is good
enough to use the 'absolute-zero' approximation in which the states are
supposed to be filled up to the Fermi level and empty above it. If the
neutral level φ_0 happens to be above the Fermi level E_F as shown in Fig.
2.5, the surface states contain a net positive charge and Q_d must therefore
be smaller than if surface states were absent. This means that the width w
of the depletion region will be correspondingly reduced, and the amount of
band bending (proportional to w^2 according to eqn A.1) will also be decreased
The barrier height φ_b is equal to the diffusion voltage or band bending
V_{d0} plus ξ (see Fig. 2.2(d)) so φ_b will be reduced.

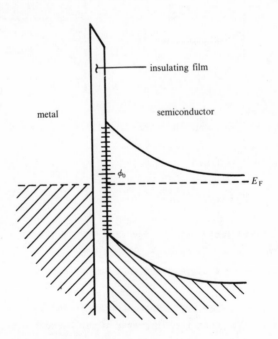

FIG. 2.5. Metal—semiconductor contact with surface states.

The reduction in φ_b has the effect of pushing φ_0 towards E_F; i.e. it tends to reduce the positive charge in the surface states. On the other hand, if φ_0 happens to be below E_F, Q_{ss} is negative and Q_d must be greater than if surface states were absent. This means that w and φ_b will both be increased and φ_0 will be pulled up towards E_F.

The surface states therefore behave like a negative-feedback loop, the error signal of which is the deviation of φ_0 from E_F. The 'gain' in this feedback loop is proportional to the density of surface states per unit of energy, since this is what determines Q_{ss} for a given departure of φ_0 from E_F. If the density of surface states becomes very large, the error signal will be very small and $\varphi_0 \approx E_F$. It is usual in the literature to measure φ_0 from the top of the valence band, in which case the barrier height will be given by

$$\varphi_b \approx E_g - \varphi_0.$$

This will be called the Bardeen limit. The barrier height is said to be 'pinned' by the high density of surface states.

An alternative way of looking at the effect of surface states is to regard them as screening the semiconductor from the electric field in the insulating layer, so that the amount of charge in the depletion region (and therefore the barrier height) is independent of the work function of the metal.

Another consequence of the existence of surface states is that the bands may not be horizontal near the free surface of a semiconductor even when it is not in contact with another solid. If the Fermi level does not coincide with the neutral level, there will be a net charge at the surface due to the surface states, and this will produce an electric field in the semiconductor which causes bending of the bands. If the surface charge is negative, the bands will bend upwards towards the surface, and the electron concentration at the surface will be less than that in the interior of the semiconductor. In this situation the surface is said to be depleted (assuming the semiconductor to be n-type). If the charge is positive, the bands will bend downwards, the electron concentration at the surface will exceed that in the interior, and the surface is said to be accumulated. When the bands are bent near the surface, the work function is still defined as the difference between the vacuum level and the Fermi level, but this is no longer equal to $\chi_s + \xi$. If the density of surface states is very large, the Fermi level becomes locked to the neutral level and the work function is almost independent of the donor density, so that

$\varphi_s \approx \chi_s + E_g - \varphi_0$. Allen and Gobeli (1962) have observed this behaviour in silicon. The discussion in § 2.2.1, illustrated in Fig. 2.4, assumes that there are no surface states, so that $\varphi_s = \chi_s + \xi$.

2.3. GENERALIZED ANALYSIS OF THE BARDEEN MODEL

The model described in § 2.2.2, in which it is supposed that the semiconductor and metal are separated by a thin insulating layer and that there are localized states at the insulator—semiconductor interface, has the merits of being easy to analyse and of corresponding to the practical case of a semiconductor covered with a thin oxide film. It is oversimplified because it ignores any possible modification of the surface-dipole contributions to the work functions of the metal and semiconductor when they come into contact with the insulating layer, and also assumes that the interface states can be represented by point charges whereas, in practice, they are smeared out over a distance of 5—10 Å. It has, however, been widely used to interpret experimental studies of Schottky barriers and will for this reason be analysed in some detail.

2.3.1. *The flat-band barrier height*

A band diagram of a metal—semiconductor contact as postulated by Bardeen is shown in Fig. 2.6(a). The 'vacuum level' for the insulating interfacial layer may be visualized by supposing that the contact is bounded by a flat surface perpendicular to the plane of the junction (i.e. parallel to the plane of the Figure). The vacuum level is then the energy of an electron at rest outside this surface. The insulator is represented as having a well defined conduction and valence band although, for a film of about 10 Å thick, the band structure may be very different from that of the bulk. Fig. 2.6(a) shows the situation when a forward bias V is applied between semiconductor and metal. A current will, of course, flow because of the forward bias, but this will not affect the charges or energy relationships within the system. The barrier height φ_b is defined as the difference in energy between the bottom of the conduction band at the boundary of the semiconductor and the Fermi level in the metal. This is because, if the interfacial layer is very thin, electrons can readily tunnel through it, so the barrier to electron flow is determined by the maximum height of the bottom of the conduction band in the semiconductor.

There are three distinct sources of charge in the system. The first Q_m resides on the surface of the metal, the second Q_d is due to the uncompensated donor ions in the depletion region, and the third Q_{ss} is due to

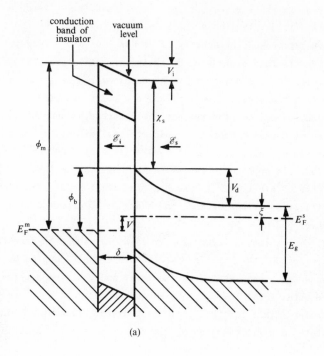

(a)

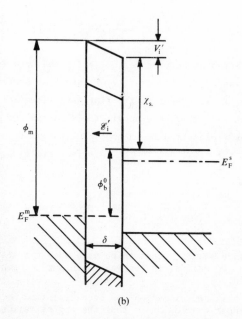

(b)

FIG. 2.6. Metal–semiconductor contact with insulating interfacial layer
(a) with arbitrary bias and (b) flat-band case.

electrons in the surface (or interface) states*. Q_m is determined by the
electric-field strength in the insulating layer, Q_d by the width of the depletion region and the density of donors, while Q_{ss} is determined by the density of interface states and by their occupation probability. Because there
is no electric field in either the metal or the semiconductor at some distance from the junction, electrical neutrality requires that $Q_m + Q_d + Q_{ss}$
= 0, and in addition the various field strengths and charge densities are
related by Gauss's theorem. A complete analysis of the problem involves a
lot of tedious algebra, but it can be simplified by supposing that sufficient bias is applied to the system to make the depletion region disappear
as in Fig. 2.6(b). This is known as the 'flat-band' situation, and the
simplification arises because now there is no electric field in the semiconductor and $Q_d = 0$.

If the insulating layer is very thin, say about 10 Å as in a good
Schottky diode, the interface states are closely coupled by tunnelling to
the conduction-band states in the metal and their population is determined
by the metal Fermi level. It is a good approximation to use the zero-
temperature limit for the Fermi—Dirac distribution; i.e. to suppose that
the states are filled up to E_F^m and empty above it.[†] Hence, by definition of
the neutral level,

$$Q'_{ss} = qD_s \ (\varphi_b^0 + \varphi_0 - E_g) \qquad (2.3)$$

where φ_b^0 is the flat-band barrier height, φ_0 is measured from the top of
the valence band, and D_s is the density of interface states per unit area
per eV. Primed quantities refer to the flat-band situation, and all charges
refer to unit area. There is no electric field in the semiconductor or in
the metal, and the field strength in the insulator is related to the charge
on the surface of the metal and in the interface states by Gauss's theorem,
so that

$$\varepsilon_i \mathscr{E}'_i = Q'_{ss} = - Q'_m$$

---

*The term 'interface states' tends to be used when the states are at the
interface between two solids, in contrast with surface states at the free
surface of a solid.

[†]The validity of this approximation has been discussed by Crowell and
Roberts (1969). Their conclusion is that the approximation is a good one
provided the density of states does not change very much over an interval
of kT/q.

where ε_i ($= \varepsilon_{ir}\varepsilon_0$) is the permittivity of the interfacial layer, and the drop in potential across the layer is given by*

$$V_i' = \delta \mathscr{E}_i' = \delta Q_{ss}'/\varepsilon_i.$$

The positive direction of $\mathscr{E}$ is taken to be from right to left in Fig. 2.6. If we assume that the dipole contributions to φ_m and χ_s are unchanged by the presence of the insulating film, we have from Fig. 2.6(b) that

$$\varphi_m = V_i' + \chi_s + \varphi_b^0$$

so that

$$\varphi_b^0 = \varphi_m - \chi_s - (\delta Q_{ss}'/\varepsilon_i) \qquad (2.4)$$

or, making use of eqn (2.3),

$$\varphi_b^0 = \varphi_m - \chi_s - \frac{q\delta D_s}{\varepsilon_i}(\varphi_b^0 + \varphi_0 - E_g) \qquad (2.4a)$$

i.e.

$$\varphi_b^0 = \gamma(\varphi_m - \chi_s) + (1-\gamma)(E_g - \varphi_0) \qquad (2.5)$$

where

$$\gamma = \frac{\varepsilon_i}{\varepsilon_i + q\delta D_s}. \qquad (2.6)$$

Eqn (2.5) was first derived by Cowley and Sze (1965) as an approximation to the zero-bias case, but they seem to have been unaware that the approximation they made in obtaining it was equivalent to neglecting Q_d. It may be observed that φ_b^0 tends to the Mott limit $\varphi_m - \chi_s$ as $D_s \to 0$ and to the Bardeen limit $E_g - \varphi_0$ as $D_s \to \infty$. Taking 20 Å as an upper limit for δ in a reasonably good Schottky diode, and assuming ε_i to be 3×10^{-11} F m^{-1} (corresponding to SiO_2), we find that D_s must approach 10^{17} eV^{-1} m^{-2} to make γ significantly less than unity.

It is possible to envisage a situation in which the surface-state den-

---

*All energies are measured in eV, so the drop in potential is numerically equal to the fall in the vacuum level across the interfacial layer.

sity is not constant but is strongly peaked about an energy φ_1 which differs from the neutral level φ_0. In this case the charge in the interface states will be given by

$$Q'_{ss} = qD_s(\varphi_b^0 + \varphi_1 - E_g) + Q_1 \qquad (2.3a)$$

where Q_1 is the charge in the surface states when they are filled up to φ_1, and D_s is the density of states near φ_1. Eqn (2.5) now becomes

$$\varphi_b^0 = \gamma\left(\varphi_m - \chi_s - \frac{\delta Q_1}{\varepsilon_i}\right) + (1-\gamma)(E_g - \varphi_1). \qquad (2.5a)$$

In the Bardeen limit, the barrier height will be pinned to the value $E_g - \varphi_1$, and the Fermi level will be close to the peak in the surface-state distribution.

2.3.2. *The field dependence of the barrier height*

If there is no interfacial layer, the barrier height is independent of any electric field which may exist inside the semiconductor. However, when there is an interfacial layer present, an electric field in the semiconductor changes the potential V_i across the layer and so modifies the barrier height. An electric field normally exists within a Schottky barrier, and it is important to know how this affects the height of the barrier.

If we revert to Fig. 2.6(a) where the bands are bent, there is now an electric field $\mathscr{E}_s$ in the semiconductor due to the uncompensated donors, all the charges are modified, and the electric field in the insulator is changed.

The charge in the interface states is given by

$$Q_{ss} = qD_s(\varphi_b + \varphi_0 - E_g)$$

$$= Q'_{ss} + qD_s(\varphi_b - \varphi_b^0), \qquad (2.7)$$

and Gauss's theorem now tells us that

$$\varepsilon_i \mathscr{E}_i - \varepsilon_s \mathscr{E}_{max} = Q_{ss},$$

where $\varepsilon_s \ (= \varepsilon_{sr}\varepsilon_0)$ is the permittivity of the semiconductor and $\mathscr{E}_{max}$ is the value of $\mathscr{E}_s$ at the top of the barrier.

As before, $V_i = \delta \mathscr{E}_i$ and $\varphi_m = V_i + \chi_s + \varphi_b$, so that

$$\varphi_b = \varphi_m - \chi_s - \frac{\delta}{\varepsilon_i}(Q_{ss} + \varepsilon_s \mathcal{E}_{max}) .$$ (2.8)

Combining eqn (2.8) with eqns (2.4) and (2.7) gives

$$\varphi_b = \varphi_b^0 - \frac{q\delta D_s}{\varepsilon_i}(\varphi_b - \varphi_b^0) - \frac{\delta\varepsilon_s}{\varepsilon_i}\mathcal{E}_{max}$$

or

$$\varphi_b = \varphi_b^0 - \alpha\mathcal{E}_{max}$$ (2.9)

where

$$\alpha = \frac{\delta\varepsilon_s}{\varepsilon_i + q\delta D_s} .$$ (2.10)

In other words, the barrier height is reduced from the flat-band value by
an amount which is proportional to the maximum electric field in the semiconductor. This result is true for any value of applied bias, provided
that the density of interface states D_s remains constant over the range of
energy involved. This can only be expected to be true if the reduction in
φ_b is less than about 0.1 eV.

There are other possible causes of field dependence of the barrier height
besides the presence of an interfacial layer. Apart from the image-force
lowering of the barrier to be discussed in § 2.4, there is also a possible
effect due to the penetration of the wave functions of the metal electrons
into the semiconductor, which is discussed in § 2.5. The field dependence
of the barrier height is very important in connection with the reverse
current/voltage characteristics, as will be discussed in § 3.6.1.

2.3.3. p-type semiconductors

The case of a p-type semiconductor for which φ_s exceeds φ_m is analogous to the n-type case discussed in § 2.3.1. Reference to Fig. 2.7 shows
that the flat-band barrier height φ_{bp}^0 is given by

$$\varphi_m + V_i'' = \chi_s + E_g - \varphi_{bp}^0$$

where

$$V_i'' = -\delta\mathcal{E}_i'' = \frac{q\delta Ds}{\varepsilon_i}(\varphi_{bp}^0 - \varphi_0) .$$

The positive direction of $\mathscr{E}$ is taken to be from right to left as before. Eqn (2.4a) is now replaced by

$$\varphi_{bp}^{0} = E_g + \chi_s - \varphi_m - \frac{q\delta D_S}{\varepsilon_i}(\varphi_{bp}^{0} - \varphi_0)$$

so that

$$\varphi_{bp}^{0} = \gamma(E_g + \chi_s - \varphi_m) + (1-\gamma)\varphi_0, \tag{2.11}$$

where γ is given by eqn (2.6). Combining eqns (2.11) and (2.5) gives

$$\varphi_{bn}^{0} + \varphi_{bp}^{0} = E_g \tag{2.12}$$

where φ_{bn}^{0} is the flat-band barrier height between the same metal and the same semiconductor doped to make it n-type. (Eqn (2.5) refers to the n-type case and so gives φ_{bn}^{0}.) Eqn (2.12) is only valid if δ, ε_i, and D_s are the same in each case, so that γ is unaltered. This is a reasonable assumption if the semiconductor surface is prepared in the same way in both cases since the surface properties are not influenced by the bulk impurities.

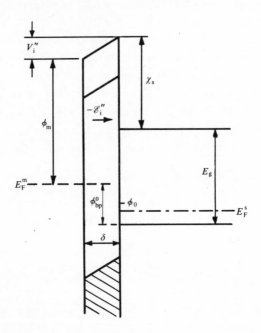

FIG. 2.7. p-type semiconductor contact with interfacial layer (flat-band case).

As with eqn (2.9), the field dependence of the barrier can be shown to be given by

$$\varphi_{bp} = \varphi_{bp}^{0} + \alpha \mathscr{E}_{max} \qquad (2.13)$$

where α is defined by eqn (2.10). Since the positive direction of $\mathscr{E}$ is taken to be from right to left, $\mathscr{E}_{max}$ is negative for a Schottky barrier on p-type material. It is therefore more appropriate to write eqn (2.13) in the form

$$\varphi_{bp} = \varphi_{bp}^{0} - \alpha |\mathscr{E}_{max}| \qquad (2.13a)$$

which is of the same form as eqn (2.9).

2.3.4. *The bias dependence of the barrier height*

Eqn (2.9) is not very useful as it stands because the electric field in the barrier is not usually known explicitly. For most purposes it is more desirable to know how the barrier height varies with bias voltage and doping level, so that we need to express $\mathscr{E}_{max}$ in terms of these parameters. This is easily done if we adopt the depletion approximation, according to which the electron density in the conduction band falls abruptly from its bulk value to a value which is negligible compared with the donor density N_d. This approximation is equivalent to supposing that the charge density rises abruptly from zero to the value qN_d at the edge of the depletion region. If N_d is constant, the electric-field strength will increase linearly with distance from the edge of the depletion region in accordance with Gauss's theorem as in Fig. 2.3(a). The field strength at the surface will be given by $\mathscr{E}_{max} = qN_d w/\varepsilon_s$, where w is the width of the depletion region, and the average field strength will be $\tfrac{1}{2}\mathscr{E}_{max}$. The difference in potential across the depletion region (i.e. the diffusion potential) will be given by

$$V_d = \tfrac{1}{2}\mathscr{E}_{max} w = \frac{\varepsilon_s \mathscr{E}_{max}^2}{2qN_d}$$

so that

$$\mathscr{E}_{max} = (2qN_d V_d/\varepsilon_s)^{\tfrac{1}{2}}. \qquad (2.14)$$

If the effect of the transition region in which the electron density falls gradually from its bulk value to a value negligible compared with N_d (sometimes incorrectly referred to as the 'reserve' layer) is properly taken into account (see Appendix B), eqn (2.14) must be modified to

$$\mathcal{E}_{max} = (2qN_d/\varepsilon_s)^{\frac{1}{2}}\left(V_d-\frac{kT}{q}\right)^{\frac{1}{2}}. \tag{2.15}$$

An expression for the barrier height in terms of the diffusion potential can be obtained by combining eqn (2.15) with eqns (2.5) and (2.9) to give

$$\varphi_b = \varphi_b^0 - \alpha(2qN_d/\varepsilon_s)^{\frac{1}{2}}\left(V_d-\frac{kT}{q}\right)^{\frac{1}{2}} \tag{2.16a}$$

$$= \gamma(\varphi_m-\chi_s) + (1-\gamma)(E_g-\varphi_0) - \alpha(2qN_d/\varepsilon_s)^{\frac{1}{2}}\left(V_d-\frac{kT}{q}\right)^{\frac{1}{2}}. \tag{2.16b}$$

Finally, an explicit expression for φ_b as a function of the bias voltage V can be obtained by inserting $V_d = \varphi_b - V - \xi$ into eqn (2.16a) and solving the resulting quadratic to give

$$\varphi_b = \varphi_b^0 + \frac{\varphi_1}{2} - \left\{\varphi_1\left(\varphi_b^0+\frac{\varphi_1}{4} - V-\xi-\frac{kT}{q}\right)\right\}^{\frac{1}{2}} \tag{2.17}$$

where $\varphi_1 = 2\alpha^2qN_d/\varepsilon_s$. The negative sign must be taken in front of the radical because φ_b is obviously less than φ_b^0. The zero-bias barrier height φ_{b0} is obtained by putting $V = 0$ in eqn (2.17).

Eqn (2.16) shows that φ_b should decrease with increasing N_d because of the increasing field in the barrier. For an interfacial layer not more than 20 Å thick, α is essentially equal to $\delta\varepsilon_s/\varepsilon_i$ unless D_s exceeds about 10^{17} eV^{-1} m^{-2}, so that α would not exceed about 60 Å if $\varepsilon_i \approx \varepsilon_s/3$. For $V_d \approx 0.5$ V and moderate doping ($N_d \lesssim 10^{22}$ m^{-3}), the lowering of the barrier due to the electric field should not exceed about 0.02 V. The effect would be exaggerated by heavier doping or a thicker interfacial layer, but would be reduced by a high surface-state density. It may be much more pronounced under reverse bias because of the increase in V_d.

2.3.5. *The penetration of the field into the metal*

Various authors (Crowell, Shore, and LaBate 1965, Perlman 1969, Kumar 1970) have taken into account the fact that the charge on the surface of the metal is only confined within the Thomas–Fermi screening distance r_0 of the surface (see, for example, Mott and Jones 1936). As a result, the electric field penetrates slightly into the metal, and there is a difference in potential between the surface and the interior of the metal of magnitude $Q_m r_0/\varepsilon_0$, where Q_m is the charge on the metal. In the case of an ideal metal—semiconductor contact with no interfacial layer and no interface states, it is easy to show that the effect of this is to reduce the height of the barrier by a fraction $2\varepsilon_s r_0/\varepsilon_0 w$, where w is the width of the depletion layer. For most metals, r_0 is about 0.5 Å and w is usually at least 1000 Å unless the donor density is very high, so for an ideal contact the effect of penetration of the field into the metal is in most cases negligible.

If there is an insulating layer between metal and semiconductor, the effect of field penetration, although still small, may be comparable with the drop in potential across the layer and may increase the field dependence of the barrier height. Eimers and Stevens (1971) have shown that the effect is simply to replace δ by $\delta + (r_0\varepsilon_i/\varepsilon_0)$ in the expressions for γ (eqn (2.6)) and α (eqn (2.10)). For chemically polished semiconductor surfaces, δ will normally be at least 20 times r_0, but the factor $\varepsilon_i/\varepsilon_0$ may be approximately equal to four. The effect of field penetration is comparable with the effect of the uncertainty in δ and ε_i.

Taking the field penetration into the metal into account is equivalent to making a partial calculation of the modification in the surface-dipole contribution to the metal's work function when it is brought into contact with the semiconductor. There are other contributions which have so far been ignored (see § 2.5). There is little point in considering just one of these contributions in isolation, particularly when its effect is so small, so the effect of field penetration into the metal will henceforth be disregarded. One must also be wary about using macroscopic concepts such as dielectric constant on an atomic scale. The dielectric constant of a material determines the average electric field over a distance of several lattice spacings, and one should not suppose, as is usually done, that the permittivity changes abruptly from ε_0 to ε_s at the metal—semiconductor boundary.

2.4. IMAGE-FORCE LOWERING OF THE BARRIER

Before we compare the predictions of theory and experiment, we must take into account the image force between an electron and the surface of the metal. In doing so, we shall assume there is no interfacial layer present.

When an electron approaches a metal, the requirement that the electric field must be perpendicular to the surface enables the electric field to be calculated as if there were a positive charge of magnitude q located at the mirror image of the electron with respect to the surface of the metal. Therefore when the electron is at a distance x from the surface of the metal it experiences a force $q^2/4\pi\varepsilon_s'(2x)^2 = q^2/16\pi\varepsilon_s'x^2$ attracting it towards the surface of the metal. Because of this attractive force the electron has a negative potential energy $-qV_I$ relative to that of an electron at infinity, where

$$V_I = \frac{q}{16\pi\varepsilon_s'} \int_x^\infty \frac{dx}{x^2} = \frac{q}{16\pi\varepsilon_s'x}$$

Following Sze, Crowell, and Kahng (1964), we have written ε_s' for the permittivity of the semiconductor because the electron approaches the metal with the thermal velocity ($\sim 10^5$ m s^{-1}), and one might expect that there is not enough time for the semiconductor to become fully polarized by the electric field, so that ε_s' should be the high-frequency rather than the static permittivity.

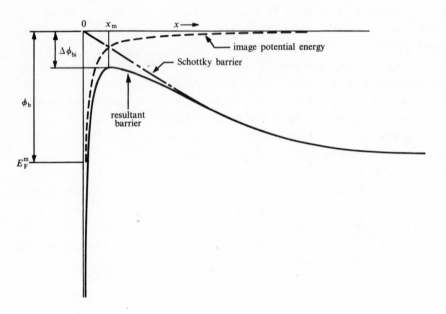

FIG. 2.8. Image-force lowering of barrier.

The image potential energy has to be added to the potential energy due to the Schottky barrier, as shown in Fig. 2.8. Since the image potential is only important near to the surface, it is a very good approximation to regard the field due to the Schottky barrier as constant with the value

$\mathcal{E}_{max}$. The maximum potential energy occurs at a position x_m where the resultant electric field vanishes; i.e. where the field due to the image force is equal and opposite to the field in the depletion region, or

$$\frac{q}{16\pi\epsilon_s' x_m^2} = \mathcal{E}_{max}.$$

The maximum potential in the barrier is lowered by an amount

$$\Delta\varphi_{bi} = x_m \mathcal{E}_{max} + \frac{q}{16\pi\epsilon_s' x_m} = 2x_m \mathcal{E}_{max}$$

as a result of the image force. Hence

$$\Delta\varphi_{bi} = 2\mathcal{E}_{max}(q/16\pi\epsilon_s'\mathcal{E}_{max})^{\frac{1}{2}}$$

$$= 2(q\mathcal{E}_{max}/16\pi\epsilon_s')^{\frac{1}{2}}$$

Substituting for $\mathcal{E}_{max}$ from eqn (2.15) and remembering that $V_d = \varphi_b - V - \xi$, we obtain

$$\Delta\varphi_{bi} = \left\{\frac{q^3 N_d}{8\pi^2(\epsilon_s')^2\epsilon_s}\left(\varphi_b - V - \xi - \frac{kT}{q}\right)\right\}^{\frac{1}{4}} \tag{2.18a}$$

and

$$x_m = \frac{1}{4}\left(\frac{q\epsilon_s}{2\pi^2(\epsilon_s')^2 N_d}\right)^{\frac{1}{4}}\left(\varphi_b - V - \xi - \frac{kT}{q}\right)^{-\frac{1}{4}}. \tag{2.18b}$$

Sze, Crowell, and Kahng (1964) have shown from photoelectric measurements of the barrier height of silicon Schottky diodes under reverse bias that the experimental data can be well explained by taking ϵ_s' equal to $(12.0 \pm 0.5)\epsilon_0$, which is indistinguishable from the static value $\epsilon_s = 11.7\epsilon_0$. This is understandable because, for the diodes under consideration, the maximum value of x_m (corresponding to zero bias) was about 50 Å, and the time taken for an electron to travel this distance with the thermal velocity is about 5×10^{-14} s. The permittivity of silicon remains constant up to a frequency ($\approx 3 \times 10^{14}$ Hz) corresponding to the band gap and, since the inverse of this is nearly an order of magnitude shorter than the electron's transit time, the silicon should become fully polarized. With a polar compound like gallium arsenide, however, there is a small decrease in permittivity above the 'reststrahlung' frequency ($\approx 7 \times 10^{12}$ Hz), so that

ε_s' should be slightly less than ε_s. Rideout and Crowell (1970) have used the values ε_{sr} = 12.5 and ε_{sr}' = 11.0 for gallium arsenide. Fig. 2.9 shows values of the image-force lowering $\Delta\varphi_{bi}$ and of the position of the maximum x_m for various values of the diffusion potential V_d, donor density N_d, and dielectric constant ε_{sr}. For this purpose ε_{sr}' has been taken equal to ε_{sr}.

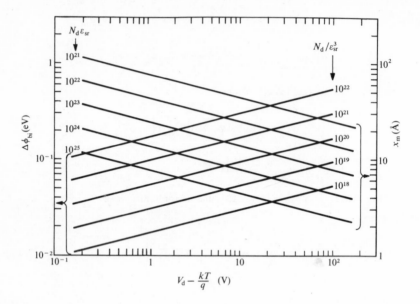

FIG. 2.9. Image-force lowering ($\Delta\varphi_{bi}$) and distance of maximum of barrier from metal (x_m) as a function of band bending for various combinations of N_d and ε_{sr}.

The effect of the image force is that the barrier which an electron has to surmount in passing from the metal into the semiconductor is lowered by an amount $\Delta\varphi_{bi}$. The image-force lowering differs from the other contributions to φ_b in that it arises from the field produced by the particular electron under consideration and is absent if there is no electron in the conduction band near the top of the barrier. On the other hand, contributions to φ_b from the work-function difference, surface-state charge etc. are present whether or not there is an electron near the top of the barrier. *We shall use φ_b to denote the barrier height arising from the latter causes*, and will denote the image-force lowering explicitly by the quantity $\Delta\varphi_{bi}$. Measurements of the barrier height which depend on the movement of conduction electrons from metal to semiconductor or vice versa yield the quantity

$\varphi_b - \Delta\varphi_{bi}$, whereas measurements which depend on the space charge in the depletion region (e.g. capacitance measurements) give φ_b without the effect of the image force.

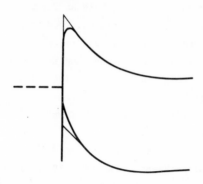

FIG. 2.10. Image-force effects in conduction and valence bands.

Like electrons, holes are attracted to the metal by an image force. However, one must remember that the energy of a hole is measured *downwards* from the top of the valence band, so that the effect of the image force is to bend the valence band *upwards* near to the surface of the metal as shown in Fig. 2.10. There is therefore no maximum in the top of the valence band as there is in the conduction band, and the energy gap is reduced close to the surface of the metal due to the effect of the image force.

2.5. INTIMATE CONTACTS

Metal—semiconductor contacts are sometimes made by cleaving the semiconductor in an ultra-high vacuum ($< 10^{-8}$ Torr) so as to create a fresh surface and then immediately evaporating the metal. The contact is thus formed before there is time for the surface of the semiconductor to become contaminated by the residual gases in the vacuum chamber, and the resulting junction is ideal from a chemical point of view because of the lack of any insulating layer between the metal and the semiconductor. There may, however, be some physical damage to the semiconductor surface as a result of the cleaving process.

Intimate contacts are more difficult to understand than those in which the metal and semiconductor are separated by an insulating interfacial layer. This is because, although the electrons can tunnel relatively easily through the interfacial layer, the latter nevertheless isolates the metal

from the semiconductor to such an extent that it is a reasonable approximation to regard the interface states as a property of the particular semiconductor–insulator combination and to ignore any modification in the
surface-dipole contributions to the work function of the metal and electron affinity of the semiconductor. With an intimate contact, however, the
electron wave functions are very seriously perturbed in the neighbourhood
of the metal–semiconductor boundary, and it is no longer possible to talk
about interface states which depend only on the semiconductor or to ignore
the modifications in the surface dipoles. The metal–semiconductor combination has to be thought of as a single system, and some attempt must be made
to calculate the change in the electron distribution which takes place
close to the boundary.

The most extensive discussion of the intimate metal–semiconductor junction seems to be that of Bennett and Duke (1967) who considered the pertur-
bations in the electron concentration at the boundary between the metal and
the semiconductor due to the abrupt change in the crystal potential. The
various forms of perturbation that can take place are shown as A, B, and C
in Fig. 2.11, which omits the band bending for clarity, and can be summarized as follows:

(a) It is possible for the wave functions of those conduction elec-

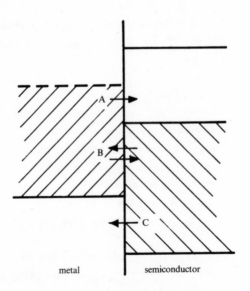

metal   semiconductor

FIG. 2.11. Redistribution of electrons in an intimate contact.

trons in the metal with energies corresponding to the forbidden gap
in the semiconductor to penetrate into the semiconductor in the form
of exponentially damped evanescent waves as was first proposed by
Heine (1965). The exponential tails of these wave functions represent a transfer of negative charge from the metal to semiconductor
as represented by the arrow A.

(b) In the case of those metal electrons which coincide in energy with
the valence band of the semiconductor, there may be a transfer of
electrons in either direction. This transfer of electrons is due
partly to the Thomas—Fermi penetration of the electric field causing
band bending in the metal, and partly due to interactions between
the electrons (the so-called correlation and exchange interactions).
Since the net direction of transfer depends on the particular circumstances, this process is represented by the two arrows B.

(c) Those electrons in the valence band of the semiconductor which have
energies below the bottom of the conduction band in the metal can
penetrate into the metal by a process similar to (a). This penetration is denoted by arrow C.

Each of the above processes results in a displacement of charge and contributes towards the surface dipole. In addition, there is the effect of the
uncompensated donors (in the case of an n-type semiconductor) in the depletion layer. Although the charge on the donors together with the balancing
charge on the surface of the metal also consistutes a dipole layer, it
differs from the above in that it extends over a distance of about 1000 Å,
whereas the others are confined to within a few ångströms of the surface.
For this reason, we do not regard the charge in the depletion layer as
forming part of the surface dipole. That part of the dipole which arises
because the balancing electrons on the surface of the metal are only confined within the Thomas—Fermi screening distance of the surface represents
the effect of penetration of the electric field into the metal, as was
discussed in § 2.3.5.

Although all the essential features of this model were discussed by
Bennett and Duke, the computational difficulties are so enormous that they
were unable to make any quantitative predictions of barrier heights of
actual metal—semiconductor combinations. One consequence of their work is
that the concept of localized surface states is no longer applicable, the
surface dipoles being a result of the modification in the electron conce-
trations close to the boundary. However, Heine (1965) has pointed out that
the tails of the metal conduction-electron wave functions (process A in

Fig. 2.11) can be thought of as replacing the Shockley–Tamm surface states, in the sense that the former would change smoothly into the latter if the metal and semiconductor were separated. Heine has discussed the effect of these tails on the barrier height, but only in a qualitative fashion. Pellegrini (1973, 1974) has attempted a detailed quantum-mechanical calculation of the effect of the tails on the barrier height but has neglected processes B and C in Fig. 2.11.

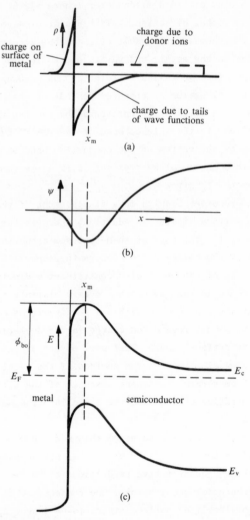

FIG. 2.12. Intimate contact for $\varphi_m < \varphi_s$. (φ_s is the work function of the semiconductor as modified by surface states.) (a) charge density, (b) electrostatic potential, and (c) band diagram. There is zero net charge either side of the plane $x = x_m$.

If the density of surface states and the position of the neutral level φ_0 at the surface of the free semiconductor are such that φ_s is greater than φ_m, as is sometimes the case*, the charge on the surface of the metal will be positive. In this case the charge distributions resulting from the Thomas—Fermi penetration of the field into the metal, the tails of the metal conduction-electron wave functions, and the uncompensated donors will be as shown in Fig. 2.12(a) leading to a band diagram as in Fig. 2.12(c). (The discontinuity at the actual boundary arises partly from the volume contributions to the work functions and partly from the surface dipoles associated with the neglected processes B and C in Fig. 2.11. In reality these will not give a perfectly abrupt discontinuity and the model is, in this sense, an idealized one.) According to this model, the total charge in the tails of the wave functions must exceed that due to the uncompensated donors in order to satisfy charge neutrality. There will then be a value of $x(=x_m)$ such that there is zero net charge to the right of the plane $x = x_m$. The electric field must therefore be zero and the potential barrier go through a maximum at $x = x_m$, as shown in the Figure. The precise position of the maximum depends on the electric field in the barrier, and this has been postulated by some authors (see § 3.6.1) as a possible cause of field dependence of φ_b in otherwise ideal barriers. This is the case discussed by Heine and Pellegrini. Neither of these authors has taken into account the effect of the band bending in the semiconductor on the precise form of the evanescent tails, nor have they taken into account the processes represented by B and C in Fig. 2.11. The problem is extremely complicated, and a completely self-consistent calculation has yet to be made. The validity of the concept of a 'neutral level' (§ 2.1.3) in the presence of the evanescent tails has also not been properly discussed, though Yndurain (1971) has shown theoretically that the presence of these tails is accompanied by a reduction in the number of states in the valence band. He has also shown that the density of the tail states depends on the Fermi energy in the metal, so that the barrier height is not a monotonic function of φ_m. According to Tejedor, Flores, and Louis (1977), the neutral level also is not constant for a particular semiconductor but depends on the metal.

A rather different approach to the problem of the intimate contact has recently been taken by Inkson (1973, 1974), who has developed the idea of

---

*We are here referring to the work function modified by the presence of surface states, as discussed on p. 27. The discussion in § 2.2.1 refers to the case where there are no surface states.

the reduction in the band gap of the semiconductor near the surface of the metal which results from the opposite effects of image forces on the conduction and valence bands as discussed at the end of § 2.4. According to Inkson the band gap actually disappears at the interface, so that the semiconductor behaves like a semimetal, and the barrier height is determined by the energy at which the top of the valence band and bottom of the conduction band coalesce. This energy must coincide with the Fermi level in the metal. However, Inkson's ideas have not yet been worked out quantitatively, although they have recently received experimental support from the work of Gregory and Spicer (1975) on gallium arsenide coated with caesium.

2.6. METHODS OF MEASUREMENT OF BARRIER HEIGHTS

2.6.1. *From J/V characteristics*

It will be shown in Chapter 3 that Schottky diodes made from high-mobility semiconductors possess J/V characteristics given by the thermionic-emission theory, provided the forward bias is not too large. According to this theory,

$$J = J_0\{\exp(qV/kT)-1\} \tag{2.19}$$

where

$$J_0 = A^{**}T^2 \exp\{-q(\varphi_b-\Delta\varphi_{bi})/kT\}.$$

For convenience, we shall call $\varphi_b - \Delta\varphi_{bi}$ the effective barrier height φ_e. A^{**} is the Richardson constant modified to take into account the effective mass of electrons in the semiconductor, quantum-mechanical reflection of those electrons which are able to negotiate the barrier, and phonon scattering of electrons between the top of the barrier (as determined by the image force) and the surface of the metal. The factors which determine A^{**} are discussed in § 3.2.7..

It will be seen in Chapter 3 that in practice diodes never satisfy the ideal equation (2.19) exactly, but always the modified equation

$$J = J_0 \exp(qV/nkT)\{1-\exp(-qV/kT)\} \tag{2.20}$$

where n (which may depend on temperature) is approximately independent of V and is greater than unity. There are many possible reasons why n should exceed unity, the most common being a bias dependence of φ_b and $\Delta\varphi_{bi}$. For values of V greater than $3kT/q$, eqn (2.20) can be written in the simpler form

$$J = J_0 \exp(qV/nkT) \qquad (2.20a)$$

so that a plot of $\ln J$ against V in the forward direction should give a straight line except for the region where $V < 3kT/q$. The advantage of retaining the more exact form of eqn (2.20) is that a plot of $\ln [J/\{1-\exp(-qV/kT)\}]$ against V should give a straight line even for $V < 3kT/q$. The intersection of this straight line with the vertical axis gives the value of $\ln J_0$. There are three ways in which the barrier height can be deduced from the data:

(a) If A^{**} is known, the value of J_0 immediately gives $\varphi_e = \varphi_b - \Delta\varphi_{bi}$. Since an error of a factor of two in A^{**} gives rise to an error of less than kT/q in φ_e, it is not necessary to know A^{**} very accurately. The value of the barrier height found by extrapolating a logarithmic plot of $\ln [J/\{1-\exp(-qV/kT)\}]$ to $V = 0$ is the value of φ_e for zero bias (see eqn (3.12)], which we shall write as $\varphi_{e0} = \varphi_{b0} - (\Delta\varphi_{bi})_0$, where $(\Delta\varphi_{bi})_0$ is the image-force lowering given by eqn (2.18a) with $V = 0$.

(b) If A^{**} is not known, one may measure the forward J/V characteristics over a range of temperatures and hence find J_0 as a function of T. A plot of $\ln(J_0/T^2)$ against T^{-1} should give a straight line of slope $-q\varphi_{e0}/k$ and intercept on the vertical axis equal to $\ln A^{**}$. The barrier height is generally a decreasing function of temperature because the expansion of the lattice causes changes in the work functions and other parameters which determine φ_{b0}. To a first approximation we may write $\varphi_{e0}(T) = \varphi_{e0}(0) - bT$, in which case the slope of the plot of $\ln (J_0/T^2)$ against T^{-1} is $-q\varphi_{e0}(0)/k$ and the intercept is $\ln A^{**} + (qb/k)$. This method therefore gives the barrier height at 0 K.

(c) According to eqn (2.19), the reverse current should saturate at the value J_0 if $V < -3kT/q$. In practice this behaviour is rarely found, and there are several reasons (discussed in Chapter 3) why the reverse current does not saturate. By plotting $\ln J$ against some suitable function of the reverse-bias voltage (e.g. $(-V)^{\frac{1}{4}}$ if the lack of saturation happens to be due to image-force lowering), it is possible to extrapolate the curve to $V = 0$ to find J_0. If A^{**}

is known, φ_{e0} can be found directly from J_0 as in (a). If A^{**} is not
known, a knowledge of J_0 as a function of temperature enables
$\varphi_{e0}(0)$ and A^{**} to be determined as in (b).

It should be stressed that the determination of barrier heights from J/V
characteristics is only reliable if the forward plot of $\ln J$ against V is a
good straight line with a low value of n, say $n < 1.1$. For large values of
n, or non-linear plots of $\ln J$ against V, the diode is far from ideal
probably due to the presence of a thick interfacial layer or to recombination in the depletion region, and the barrier height is not clearly defined.

2.6.2. *From photoelectric measurements*

If radiation with quantum energy exceeding $\varphi_b - \Delta\varphi_{bi}$ is incident on a metal
in contact with a semiconductor, electrons excited from the Fermi level of
the metal will have sufficient energy to cross into the semiconductor, and
a photovoltaic e.m.f. will be developed causing a current to flow in an
external circuit. According to Fowler (1931), the photocurrent γ per
absorbed photon of energy $h\nu$ is given by

$$(\gamma/T^2) = B\left[\frac{\pi^2}{6} + \frac{\mu^2}{2} - \left(e^{-\mu} - \frac{e^{-2\mu}}{2^2} + \frac{e^{-3\mu}}{3^2} - \dots\right)\right]$$

where B is a constant and $\mu = h(\nu-\nu_0)/kT$. The threshold energy $h\nu_0$ is equal
to $\varphi_b - \Delta\varphi_{bi}$. For $\mu \gg 1$, γ is proportional to $\{h(\nu-\nu_0)\}^2$, so a plot of $\gamma^{\frac{1}{2}}$
against $h\nu$ should give a straight line the intercept of which on the $h\nu$
axis is equal to $h\nu_0$. This again gives the barrier height reduced by the
image-force term $\Delta\varphi_{bi}$. A detailed discussion of the photoelectric method
has recently been given by Anderson, Crowell, and Kao (1975).

2.6.3. *From capacitance measurements*

It will be seen in Chapter 4 that, subject to certain precautions being
taken in the measurement and provided the diode is nearly ideal and the
semiconductor has a uniform donor concentration, the differential capacitance C $(=\frac{dQ}{dV})$ under reverse bias V_r is given for non-degenerate semiconductors by

$$C = S(qN_d\varepsilon_s/2)^{\frac{1}{2}}\left(\varphi_b - \xi + V_r - \frac{kT}{q}\right)^{-\frac{1}{2}}$$

where S is the area of the contact.

Hence

$$c^{-2} = (2/s^2 q N_d \varepsilon_s)\left(\varphi_b - \xi + V_r - \frac{kT}{q}\right).$$

If φ_b is independent of V_r (i.e. if there is no appreciable interfacial layer), a plot of c^{-2} against V_r should give a straight line with an intercept $-V_I$ on the horizontal axis equal to $-\left(\varphi_b - \xi - \frac{kT}{q}\right)$. The barrier height is then given by

$$\varphi_b = V_I + \xi + \frac{kT}{q}. \qquad (2.21)$$

If, however, there is an appreciable interfacial layer, so that φ_b depends on V_r, it has been shown by Cowley (1966) that the intercept of the c^{-2} against V_r plot when inserted into eqn (2.21) yields the quantity $\varphi_b^0 + (\varphi_1/4)$, where φ_b^0 is the flat-band barrier height and $\varphi_1 = 2\alpha^2 q N_d / \varepsilon_s$ (see § 2.3.4 and § 4.2). Since in all practical cases φ_1 is less than φ_b^0 by about two orders of magnitude, the c/V method essentially gives the flat-band barrier height. Furthermore, because the differential capacitance is determined by the width of the depletion region which depends only on the diffusion voltage and the donor density, the barrier height given by eqn (2.21) does not include the image-force lowering $\Delta\varphi_{bi}$. A comprehensive discussion of the various sources of error in the determination of barrier heights from c/V characteristics has been given by Goodman (1963). In the case of low barrier heights (say less than 0.5 V), the parallel conductance of the diode may be so large that it is not possible to make capacitance measurements unless the diode is cooled.

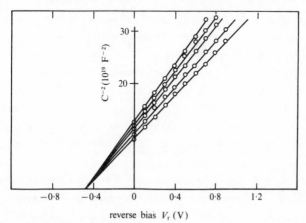

FIG. 2.13. Plots of c^{-2} against V_r for five gold–silicon Schottky diodes made simultaneously by evaporation onto a single slice of silicon (Turner and Rhoderick 1968).

Computer programs for the numerical determination of barrier heights from the J/V and C/V characteristics and by the photoelectric method have recently been published by Nguyen, Lepley, Nadeau, and Ravelet (1975). The reliability of the three methods can be good in the sense that measurements on a number of diodes manufactured simultaneously by evaporation of the metal onto a single semiconductor slice are generally in good agreement; for example, Fig. 2.13 shows plots of C^{-2} against V_r for gold—silicon diodes made by simultaneous evaporation on to a single slice of silicon. However, agreement between measurements on diodes made on separate slices at different times is usually less good and reflects the reproducibility of the fabrication procedure. Generally, for diodes which have been made on separate occasions but using the same fabrication procedure, measurements made by any one of the three methods can be reproducible to about 2%. If there is a pronounced bias dependence of the barrier height, the capacitance method is liable to give results which differ significantly from those obtained by the J/V or photoelectric methods. This is the case when there is a fairly thick interfacial film, as was pointed out by Cowley (1966) and by Card and Rhoderick (1971a). Determinations of φ_b from the reverse J/V characteristics can only be regarded as reliable if a guard-ring is used, as will be discussed in § 3.6.2.

2.7. BARRIER-HEIGHT MEASUREMENTS ON ETCHED SURFACES

2.7.1. *Silicon*

Because of their practical importance, there is more information available about contacts on etched surfaces of silicon than on any other semiconductor. An etched surface is one prepared by the usual technological processes of cutting, polishing and chemical etching. The evaporation of the metal is then usually carried out in a conventional vacuum system having a background pressure of around 10^{-6} Torr. It is well known that silicon surfaces prepared in this way are covered with a thin insulating film, usually of SiO_2. The thickness of this oxide layer can be measured by ellipsometry (Archer 1962) and values between 10 Å and 20 Å are usually obtained, depending on the method of surface preparation.

The effect of the method of surface preparation was first described in some detail by Turner and Rhoderick (1968), who made an extensive study of the effects of different types of etching techniques on the barrier heights obtained by evaporating various metals onto (111) silicon surfaces. The barrier heights were deduced from the J/V characteristics, both from

measurements of J_0 (assuming a value of A^{**} as discussed in § 2.6.1(a))
and from the temperature variation of J_0 as in § 2.6.1(b), and also from
the C/V characteristics. They found general agreement between the first
and last methods within 0.02 eV, but the temperature-variation method
consistently gave results about 0.05 eV higher. As was pointed out in
§ 2.6.1, the temperature-variation method gives the barrier height at 0 K,
so the measurements imply that φ_{e0} is a decreasing function of temperature.
For the most heavily doped samples ($N_d = 5 \times 10^{21}$ m^{-3}), the image-force
lowering amounted to about 0.02 eV for a gold contact, and because this
was not greater than the experimental error and the values obtained from
the C/V characteristics were not consistently larger than those obtained
from the J/V plots, the image-force lowering was ignored so that the
measurements gave φ_b^0 or φ_{b0}, which were indistinguishable within experimental accuracy.

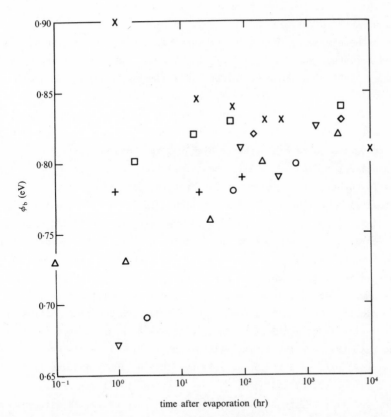

FIG. 2.14. Ageing of gold—silicon Schottky diodes. The different symbols
refer to different methods of surface preparation. (Turner and Rhoderick
1968.)

The results showed that the barrier heights depended on the method of preparation of the silicon surface and that the contacts aged with time. Fig. 2.14 shows the variation of the barrier heights with time for gold contacts on silicon after using various methods of surface preparation. The barrier heights settled down to a steady value after 100 to 1000 hours, the final values being relatively insensitive to the method of surface preparation. It is reasonable to suppose that this ageing was connected with some slow change in the interfacial layer, and the time scale suggests the migration of charged ions, as often occurs in MOS transistors. During this migration the ions, together with the compensating charges on the surface of the metal, would give rise to a dipole which would modify the barrier height in the same way as surface states. This explanation is consistent with the fact that the ageing process can be accelerated by heating, and also explains why no ageing was observed in the case of 'intimate' contacts, which should be free from any oxide film (see § 2.8.1).

Measurements of barrier heights on chemically etched silicon surfaces have also been made by several other workers, notably Kahng (1963), Jäger and Kosak (1969), and Hirose, Altaf, and Arizumi (1970). Their results are generally similar to those of Turner and Rhoderick, although they did not study ageing effects. In all cases, the donor density was such that the theoretical value of the image-force lowering was less than the experimental error, and there was no tendency for barrier heights obtained from C/V characteristics to be consistently higher than those obtained from the I/V or photoelectric data. Fig. 2.15 shows data from the above sources, together with a few other results on individual metals, plotted against commonly accepted values of the metal work functions. There is a difficulty in knowing what values of the work function to use, because of the wide scatter of experimental values. Apart from the differences between the values obtained with different methods of measurement, it is found that in the case of evaporated-metal films the work function depends on the choice of substrate. The dilemma is well illustrated by gold, the work function of which was determined as 4.7 eV in a vacuum system fitted with a mercury diffusion pump (Rivière 1957) and 5.25 eV in an ultra-high-vacuum system fitted with an ion pump (Rivière 1966). The values used in Fig. 2.15 are averages of the values quoted in a review article by Rivière (1969) and are listed in Table 2.1. For nickel and gold, we have retained the lower values of 4.73 eV and 4.71 eV, respectively, because the higher values have only been obtained in ultra-high-vacuum systems. The barrier heights were obtained using a variety of methods on silicon with various donor concentrations.

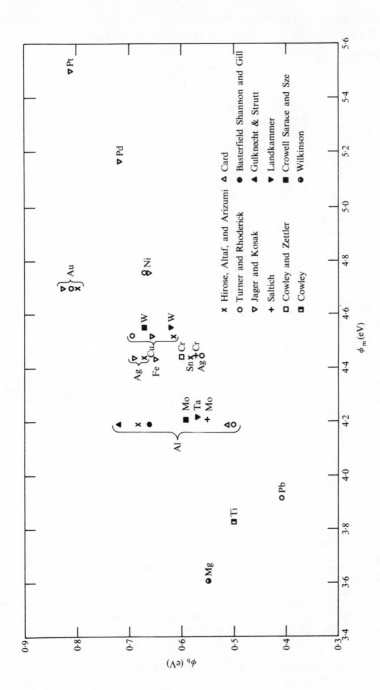

FIG. 2.15. Barrier heights on chemically etched silicon surfaces. The values of ϕ_m are average values taken from Table 2.1, except for Ni and Au, where the lower values have been used.

There appears to be no systematic dependence of barrier height on the method of measurement or on the doping level, so there is no clear distinction between φ_{b0} and φ_b^0. For this reason the barrier heights have simply been labelled φ_b.

TABLE 2.1

Work functions of metals

Element	Rivière			Thanailakis and Rasul	Average (eV)
	Contact potential (eV)	Photoelectric (eV)	Thermionic emission (eV)	Contact potential (eV)	
Na[§]					2.28
Mg	3.61				3.61
Al	4.19			4.17	4.18
K		2.22			2.22
Ca[‡]					2.86
Ti		3.57	4.1		3.83
Cr		4.44			4.44
Mn		4.08			4.08
Fe	4.16	4.65		4.58	4.46
Co		4.97		4.97	4.97
Ni[†]	{4.73 (h.v.) / 5.25 (u.h.v.)	5.10		5.10	5.15
Cu	4.51	4.70		4.55	4.59
Mo	4.21				4.21
Pd		5.17			5.17
Ag	4.44	4.42		4.41	4.42
In		3.97			3.97
Sn	4.43				4.43
Sb		4.56			4.56
Ba	2.48				2.48
Hf			3.63		3.63
Ta*	4.22	4.16	4.23		4.20
W*	4.55	4.55	4.55		4.55
Pt		5.63	5.36	5.30	5.43
Au[†]	{4.71 (h.v.) / 5.30 (u.h.v.)			5.10	5.20
Pb	3.83			4.25	4.04

The first three columns contain data from the review by Rivière (1969), and the fourth column gives the results of Thanailakis (1975) and Thanailakis and Rasul (1976).

*The data refer to evaporated films with the exception of those for Ta and W which refer to polycrystalline bulk samples.

[†]The data quoted by Rivière were obtained in conventional diffusion-pumped h.v. systems with the exception of the second row of figures for Ni and Au which were obtained in ion-pumped u.h.v. systems. The results of Thanailakis and Rasul were all obtained in an ion-pumped u.h.v. system.

[‡]The figure quoted for Ca is the average of data given in the Handbook of Chemistry and Physics.

[§]From Maurer, R.J. (1940), *Phys. Rev.*, **57**, 653.

It is apparent that there is no close correlation between barrier height
and work function, and that there can be a wide scatter between results for
the same metal (e.g. aluminium) obtained by different workers. Some of the
scatter is undoubtedly due to variation in the surface treatments used by
different authors and can be partially attributed to the lack of information on ageing effects. Furthermore, some workers heated the silicon to
drive off moisture while others did not. All that can be said with confidence is that metals with a high work function (e.g. gold, platinum) tend
to give large barrier heights, while metals with low work functions (e.g.
magnesium, calcium) tend to give small barrier heights. Inert metals like
gold usually give reproducible results, whereas reactive metals like aluminium show a good deal of scatter, possibly because aluminium oxidises
easily and tends to reduce the natural film of SiO_2 left on the surface of
the silicon by the etching process.

The results of Turner and Rhoderick taken in isolation (Fig. 2.16) show
a more pronounced correlation between barrier height and work function than
do all the data taken collectively. This may be because they used a stan-
dard method of surface preparation and allowed for ageing effects. Their
data can be reasonably well fitted by a straight line of slope about 0.5.
They found no tendency for more highly doped silicon to give lower barriers
than more weakly doped material, so the term $\alpha \mathscr{E}_{max}$ in eqn (2.9) can be
neglected and we may take $\varphi_{b0} \approx \varphi_b^0$. Their results taken in conjunction with
eqn (2.5) therefore indicate that $\gamma \approx 0.5$, so that $(q\delta D_s/\varepsilon_i) \approx 1$. If we

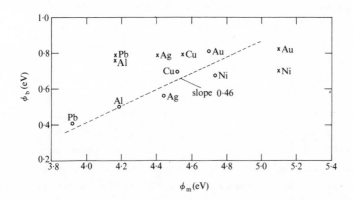

FIG. 2.16. Barrier heights on etched (○) and cleaved (X) silicon surfaces
(Turner and Rhoderick 1968). In the cases of Ni and Au, the lower values of
φ_m have been used for the etched surfaces, and the higher for the cleaved
(u.h.v.) surfaces.

take the value 15 Å for δ, which is probably correct to within 50%, we find $(D_s/\varepsilon_i) \approx 4 \times 10^{27}$ eV^{-1} F^{-1} m^{-1}. It is not possible to find D_s and ε_i separately from these experiments but, if one assumes the oxide to be SiO$_2$, the value of ε_i should not be very different from the value for bulk SiO$_2$ which has a dielectric constant of about four, since the value assumed for δ is several times the lattice spacing. Assuming $\varepsilon_i \approx 3 \times 10^{-11}$ Fm^{-1}, we find $D_s \approx 1.2 \times 10^{17}$ ev^{-1} m^{-2}, which is typical of experimental values obtained by other methods for etched silicon surfaces. The results were also consistent with a value of φ_0 of 0.31 eV above the top of the valence band. It does not seem profitable to indulge in any more elaborate attempt at comparing theory and experiment; what is significant is that the results of one particularly extensive set of experiments can be qualitatively explained in terms of the Bardeen model using reasonable values of the constants involved.

The data can be analysed in a different way by using some recent results of Kar (1975). Kar made direct measurements of the quantity $\varphi'_m - \chi'_s$ for a number of metals, where φ'_m and χ'_s are respectively the work function of the metal and the electron affinity of silicon *measured relative to the bottom of the conduction band of SiO$_2$*. These measurements were carried out by measuring the flat-band voltage of MOS capacitors as a function of oxide thickness. Separate measurements of φ'_m and χ'_s have also been made by Deal, Snow, and Mead (1968) using photoelectric methods. The agreement between the two sets of measurements is very good. The quantity $\varphi'_m - \chi'_s$ should be the same as $\varphi_m - \chi_s$ only if the dipole contributions to the work function and electron affinity are not modified when the metal and the silicon are in contact with SiO$_2$ rather than with vacuum. However, if the effect of chemical etching in the fabrication of metal—semiconductor contacts is to leave a thin layer of oxide on the surface of the silicon which we assume to be SiO$_2$, it is the quantity $\varphi'_m - \chi'_s$ which should really be used in eqn (2.5). Since Kar's measurements seem to be very reliable (he quotes errors in the range 0.03 to 0.06 eV for $\varphi'_m - \chi'_s$) and also give a direct value of the difference between φ'_m and χ'_s, we should expect to get better correlation by plotting φ_b directly against $\varphi'_m - \chi'_s$ than by using doubtful values of the work function φ_m. Fig. 2.17 shows the data for the seven metals studied by Kar. There is no better correlation than when the metal work functions are used, and it is particularly noticeable that the metals with large negative values of $\varphi'_m - \chi'_s$ (magnesium and tin) give quite substantial barrier heights. In the case of magnesium, if Kar's value of $\varphi'_m - \chi'_s$ is taken to apply to the Schottky barrier, the field across the oxide layer

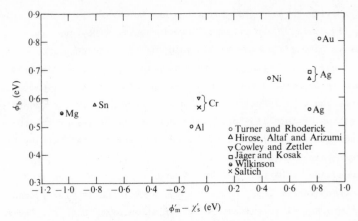

FIG. 2.17. Barrier heights on etched silicon surfaces, plotted against $\varphi'_m - \chi'_s$ ($\varphi'_m - \chi'_s$ taken from Kar 1975, except value for Ni which is taken from Deal, Snow, and Mead (1968)).

(assuming its thickness to be about 15 Å) must be about 10^9 V m^{-1}. This is very high and corresponds to a negative charge in the surface states of about 3×10^{-2} C m^{-2}. This means that the density of surface states must be around 6×10^{17} eV^{-1} m^{-2} (assuming $\varphi_0 = 0.3$ V), which is an exceptionally large value for an etched surface. It is also noticeable that the barrier height for nickel falls reasonably amongst the other data when plotted against $\varphi'_m - \chi'_s$, whereas Fig. 2.15 suggests that the adopted value of the nickel work function is much too high.

An alternative way of analysing barrier-height measurements has been suggested by Mead (1966), who proposed that φ_b should be considered as a function of the electronegativity of the metal rather than its work function. Electronegativity is a concept introduced by Pauling (1960) to describe an atom's ability to form covalent bonds with other atoms; it is only a semiquantitative concept, but can be represented by a parameter with the dimensions of energy of the order of 1 eV. It is not clear to the writer that there is any particular justification for analysing the data in this way. The only advantage seems to be that a table of electronegativities is given in Pauling's book, which avoids the difficulty of knowing what value of work function to use. On the other hand, electronegativity cannot be precisely defined and is only quoted by Pauling to within 0.1 eV; moreover, it is a property of a single atom, not of a solid, and does not take into account the surface dipole which contributes to the work function and which undoubtedly affects the barrier height when a metal makes contact with a semiconductor.

An interesting series of measurements has been made by Arizumi, Hirose, and Altaf (1968, 1969) who prepared Schottky barriers by evaporating binary alloys of pairs of the metals gold, silver, and copper onto silicon. They found that for any pair the barrier height varied linearly with alloy composition between the values corresponding to the pure metals. One might expect the work functions of alloys to vary in a linear manner with the composition of the alloy provided the band structures of the constituent metals are similar, but a review of the experimental data by Rivière (1969) shows this not to be the case.

The dependence of barrier height on donor density has been studied by several workers (Kahng 1963, Saltich 1969, Archer and Yep 1970), and in no case was there any clear evidence of a variation of φ_b with N_d. The most extensive investigation was that of Archer and Yep, who measured the height of barriers formed by evaporating gold onto etched silicon with donor densities between 10^{20} and 10^{25} m^{-3}. They measured the barrier heights by the capacitance and photoelectric methods, and corrected the latter for the effects of image-force lowering and tunnelling; they found φ_b to be independent of N_d within the experimental error of $\pm$ 0.05 eV, which implies that the zero-bias barrier height φ_{b0} must be virtually indistinguishable from the flat-band barrier height φ_b^0, so that $\varphi_b^0 - \varphi_{b0} < 0.05$ eV. Assuming the Bardeen model, we can use this result in conjunction with eqn (2.17) to find an upper limit for α. The dependence of φ_{b0} on N_d is contained in φ_1 $(= 2\alpha^2 q N_d/\varepsilon_s)$, and putting $V = 0$ in eqn (2.17) and neglecting φ_1, ξ, and kT/q in comparison with φ_b^0 gives $\varphi_{b0} \approx \varphi_b^0 - (\varphi_1 \varphi_b^0)^{\frac{1}{2}}$, so that $(\varphi_1 \varphi_b^0)^{\frac{1}{2}} < 0.05$ eV. Taking $\varphi_b^0 = 0.8$ eV, $N_d = 10^{25}$ m^{-3}, and $\varepsilon_s = 10^{-10}$ F m^{-1}, we find that

$$\alpha = \delta \varepsilon_s / (\varepsilon_i + q \delta D_s) < 3 \times 10^{-10} \text{ m}$$

or

$$(\varepsilon_i/\delta) + q D_s > 3 \times 10^9 \; \varepsilon_s = 0.3 \text{ F m}^{-2}.$$

Assuming that $\varepsilon_i \approx 4\varepsilon_0$ and $\delta \approx 15$ Å, the first term on the left is negligible so that $D_s > 2 \times 10^{18}$ eV^{-1} m^{-2}. This is an exceptionally high density of surface states for an etched surface and is more like that characteristic of a cleaved surface. Archer and Yep suggest that this unexpectedly large density of surface states throws doubt on the validity of the Bardeen model.

The temperature dependence of the barrier height has been studied by several workers with conflicting results. Crowell, Sze, and Spitzer (1964) measured the temperature dependence of gold–silicon barriers using the photoelectric method and found that φ_b was a decreasing function of temperature with a functional dependence almost identical with the temperature dependence of the energy gap. Arizumi and Hirose (1969) also found the height of gold–silicon barriers to decrease more or less linearly with increasing temperature with a coefficient of about -3×10^{-4} eV K^{-1}. On the other hand, Cowley (1970) observed that the temperature dependence of titanium–silicon barriers was much less than that of the energy gap, and Cowley and Zettler (1968) state that the barrier height of their chromium–silicon diodes was the same, within the experimental error, at 196 K and 298 K. On theoretical grounds one would expect a temperature variation of the energy gap to result in a temperature variation of φ_b because of a change in $E_g - \varphi_0$. It is also to be expected that the work function of the metal and electron affinity of the silicon would change with temperature although there do not seem to be any experimental measurements of this. As was pointed out in § 2.6.1, if φ_b is a function of temperature, a determination of barrier height from the slope of a plot of ln (J_0/T^2) against T^{-1} gives the value of φ_{e0} at 0 K.

Barriers on p-type silicon have been thoroughly investigated by Smith and Rhoderick (1971), who made measurements on six different metals. The barriers were generally lower than on n-type silicon, and in the case of gold the barrier was so low that it provided an effectively ohmic contact at room temperature and its height could only be measured by cooling to 210 K. The results can be combined with Turner and Rhoderick's data on n-type silicon to give the following results for $\varphi_{bn} + \varphi_{bp}$:

Metal	Ag	Al	Au	Cu	Ni	Pb
φ_{bn} (eV)	0.56	0.50	0.81	0.69	0.67	0.41
φ_{bp} (eV)	0.54	0.58	0.34	0.46	0.51	0.55
$\varphi_{bn}+\varphi_{bp}$ (eV)	1.10	1.08	1.15	1.15	1.18	0.96

The sum is equal, within the limits of experimental error, to the energy gap of silicon (1.1 eV) in all cases except perhaps that of lead which is slightly low.

2.7.2. *Germanium*

In view of its great importance in the early days of semiconductor technology, it is surprising how little information is available about Schottky barriers on germanium. There was an extensive study of point contacts in connection with early microwave-mixer development, but these contacts were always 'formed' in such a way that the junction was not a simple metal–semiconductor contact but possibly a p–n junction (see Shockley 1950). An extensive series of measurements was made by Jäger and Kosak (1969) who prepared Schottky barriers by evaporating a series of metals onto silicon and germanium heated to 300°C. The barrier heights were in all cases lower on germanium than on silicon by between 30 and 50 percent, but there appeared to be no simple relationship between the two. No attempt to explain their results in terms of the Bardeen model was made.

Schottky barriers on etched germanium surfaces have also been investigated by Thanailakis and Northrop (1973). Their results do not agree very well with those of Jäger and Kosak, possibly because they did not heat the germanium prior to evaporation. Thanailakis and Northrop analysed their data in terms of the Bardeen model in a manner similar to that used by Rhoderick and Turner to interpret the results on silicon. They found that their results were consistent with a density of surface states of 2×10^{17} eV^{-1} m^{-2} and a position of the neutral level 0.13 eV above the top of the valence band. In the case of aluminium contacts, they found pronounced ageing effects and departures from ideal behaviour which they attributed to the fact that aluminium oxidizes very easily and tends to reduce the GeO_2 film on the etched surface of the germanium.

2.7.3. III–V compounds

Because of its technological importance, more information is available about metal contacts on gallium arsenide than on any other semiconductor apart from silicon. A detailed study of Schottky barriers on etched surfaces of gallium arsenide was carried out by Smith (1968, 1969a) who found that the method of surface preparation was important if nearly ideal electrical characteristics were to be obtained. The best method was to etch the surface with a 5% mixture of bromine in methanol for a period of 5–10 min, after which the etch was flushed with methanol. The gallium arsenide was heated to about 120°C prior to and during the metal evaporation to drive off any adsorbed water vapour. Diodes prepared in this way gave good J/V characteristics with n values of about 1.06 (see eqn (2.20a)) and linear c^{-2} versus V plots. For the donor density used (5×10^{22} m^{-3}), the image-force lowering should be approximately 0.03 V, and the barrier heights deduced from the J/V and photoelectric measurements tended to be

slightly lower than those obtained from the C/V data by between 0.01 and
0.04 eV. The barrier heights were also found to depend on crystal orientation. The ($\bar{1}\bar{1}\bar{1}$) and (110) surfaces gave substantially the same barrier
heights, but the values obtained on the (100) surfaces were lower by
approximately 0.10 eV. There were also significant differences between the
values obtained on the (111) and ($\bar{1}\bar{1}\bar{1}$) faces which are not equivalent in
polar compounds like gallium arsenide. Kahng (1964) has also found φ_b to be
orientation dependent. The differences may arise because the density of
surface states or the electron affinity (or both) may depend on the crystal
orientation. Gatos, Moody, and Lavine (1960) have given theoretical reasons
based on a dangling-bond model why the density of surface states should be
orientation dependent in III—V compounds, and Arthur (1966) has found experimental differences between the work functions of the (111) and ($\bar{1}\bar{1}\bar{1}$)
faces of gallium arsenide.

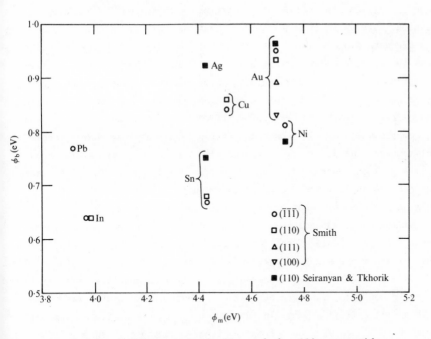

FIG. 2.18. Barrier heights on chemically etched gallium arsenide.

Smith's results are shown in Fig. 2.18, where the same work function
values as in Figs. 2.15 and 2.16 have been used and the values of φ_b are
those obtained from the C/V measurements. Considering only the ($\bar{1}\bar{1}\bar{1}$) data,

the correlation between φ_b and φ_m is seen to be poor, but it is clear that
the variation in φ_b is much less than the variation in φ_m. For what it is
worth, the best straight line, fitted by the method of least squares, has
a slope of 0.23. If the same values of ε_i and δ are assumed as in the case
of silicon (the dielectric constant of arsenic oxide is known to be about
three), the density of surface states must be about 4×10^{17} eV^{-1} m^{-2}
which is rather higher than for the etched silicon (111) surface, and φ_0
must be about 0.52 eV above the top of the valence band. Similar measurements on etched (110) surfaces have been reported by Seiranyan and Tkhorik
(1972). They find that their results can be reconciled with $D_s = 3 \times 10^{17}$
eV^{-1} m^{-2} and $\varphi_0 = 0.45$ eV.

Other measurements of barrier heights on etched surfaces of gallium
arsenide have been made by Crowell, Sarace, and Sze (1965) for tungsten
and by Kajiyama, Sakata, and Ochi (1975) for hafnium. Padovani and Sumner
(1965) and Hackam and Harrop (1972) have examined the temperature dependence of gallium-arsenide Schottky barriers with somewhat conflicting
results. Both pairs of authors found that the barrier heights deduced from
the J/V data did not agree with those obtained from C/V characteristics unless the current under forward bias was assumed to vary as $\exp\{-q\varphi_b/(T+T_0)\}$
rather than $\exp\{-(q\varphi_b/T)\}$, where T_0 is a constant with the dimensions of
temperature. After making this correction, Padovani and Sumner found that
the barrier heights of their gold–gallium arsenide diodes were independent
of temperature, whereas Hackam and Harrop found that for their nickel–
gallium arsenide diodes φ_b decreased with increasing temperature in the
same way as the energy gap. The 'T_0 effect' is discussed in § 3.9, where
the conclusion is reached that it is probably an artefact and not an intrinsic property of Schottky barriers in general.

Gallium phosphide has been extensively studied by Cowley (1966) who
showed the importance of careful control of surface conditions. He found
that, while there was good agreement between barrier heights inferred from
C/V data and those measured photoelectrically when the metal was evaporated
using an ion pump, the agreement was very poor if the system was fitted
with an oil-diffusion pump. In the latter the barrier heights obtained
from the C/V data were consistently higher than those measured photoelectrically. This could be explained in terms of the existence of an interfacial film which was much thicker with an oil-pumped system than with an
ion-pumped system (see § 4.2). In the ion-pumped system, the photoelectric
barrier heights were 1.27 ± 0.02 eV for gold and 1.44 ± 0.02 eV for
platinum on either the (111) or ($\bar{1}\bar{1}\bar{1}$) planes. (It was not known which of

these two planes the values actually referred to; the other plane was found to give barrier heights about 0.1 eV higher). Smith (1969*b*) found that consistent results could be obtained in an oil-pumped system provided the substrate was heated to 120°C during evaporation of the metal; this may hinder the formation of an interfacial layer. He obtained a value of 1.31 ± 0.02 eV for gold on the ($\bar{1}\bar{1}\bar{1}$) face of n-type gallium phosphide. For p-type material, the value was 0.88 ± 0.03 eV, and the sum of the two is in reasonable agreement with the known energy gap of gallium phosphide (2.25 eV).

Smith (1973) has also studied Schottky barriers formed by evaporating gold onto etched (110) faces of n-type indium phosphide. The barrier height had the comparatively low value of 0.40 eV which is much less than half the band gap (1.3 eV). Walpole and Nill (1971) report an investigation of gold contacts on indium arsenide. On n-type material they found the contact to be ohmic, and on p-type material the C/V characteristics suggested a barrier height of at least 0.42 eV at 77 K. This implies that the barrier height is greater than the band gap (0.40 eV at 77 K). Indium antimonide has been studied by Mead and Spitzer (1964) who found barrier heights of 0.17 eV and 0.18 eV for gold and silver, respectively, on n-type material at 77 K. Korwin-Pawlowski and Heasell (1975) also investigated contacts to n-type InSb and found that the barrier properties were very sensitive to the etch which was used. They reported a barrier height of about 0.08 eV at 77 K for both silver and gold, which is substantially lower than the values obtained by Mead and Spitzer. In both InAs and InSb, C/V measurements can only be made at low temperatures because of the very high conductance of the diodes at room temperature.

2.7.4. *Other semiconductors*

There is a limited amount of information available about metal contacts on II–VI compounds. Goodman (1964) has made a thorough study of contacts on etched single crystals of cadmium sulphide, and there is some rather sparse information about zinc telluride (Baker and Milnes 1972), zinc selenide (Livingstone, Turvey, and Allen 1973), cadmium selenide (Consigny and Madigan 1969) and zinc oxide (Neville and Mead 1970). With the exception of cadmium selenide, the barrier heights are in broad agreement with the simple Mott theory which neglects surface states. Kusaka, Matsui, and Okazaki (1974) have shown that barriers on cadmium sulphide are sensitive to the crystallographic polarity of the surface.

In spite of their importance as photodetectors, not much is known about

the IV–VI compounds (lead sulphide etc.). Nill, Walpole, Calawa, and Harman (1970) have studied contacts to lead telluride and reported that surface states are unimportant, so that the barrier heights conform approximately to the Mott limit. Since the electron affinity of lead telluride is
reported to be 4.6 eV which is greater than the work function of either
lead or tin, these metals should give ohmic contacts on n-type material
and rectifying contacts on p-type material with barrier heights greater
than the band gap. This prediction is in accordance with the C/V characteristics (Walpole and Nill 1971).

Apart from silicon and germanium, there is only scanty information
about group IV semiconductors. Gold contacts to p-type diamond have been
examined by Mead and Spitzer (1964) and by Glover (1973), who found
barrier heights of 1.35 eV and 1.73 eV, respectively. The former authors
have also looked at gold and aluminium contacts on n-type silicon carbide
and find barriers of 1.95 eV and 2.0 eV, respectively.

In view of their immense practical importance in the pre-silicon and
-germanium era, it is surprising how little is known about contacts to
cuprous oxide and selenium. Early estimates of the height of the $Cu–Cu_2O$
(p-type) barrier, prepared by oxidizing copper or by reducing cuprous
oxide, gave barrier heights of about 0.4 eV (Henisch 1957), while recent
measurements by Assimos and Trivich (1973) gave a value of 0.75 eV. Lanyon and Richardson (1971) have found the barrier between p-type selenium
and a series of metals to lie in the range 0.30 to 0.55 eV.

2.8. BARRIER HEIGHT MEASUREMENTS ON INTIMATE CONTACTS

2.8.1. *Silicon*

Contacts made by cleaving the semiconductor in an ultra-high vacuum and
then immediately evaporating the metal should be ideal from a purely chemical point of view as there is no insulating layer between the metal and the
semiconductor. Such contacts may not, however, be physically perfect; the
cleaving process is known to cause physical damage which may take a long
time to anneal out.

The first extensive application of this method to silicon was made by
Archer and Atalla (1963) and, although the vacuum used by them was only
10^{-6} Torr, the metal was evaporated so rapidly that there should have been
negligible contamination of the interface. They determined the barrier
heights from C/V measurements. Their experiments were repeated by Turner
and Rhoderick (1968) using a rather better vacuum ($\sim 10^{-8}$ Torr). The latter

also deduced their barrier heights from C/V data and found very good agreement with the results of Archer and Atalla. Similar experiments on cleaved silicon using low-work-function metals (Mg, Ca, K, and Na) were carried out by Crowell, Shore, and LaBate (1965) and by Szydlo and Poirier (1973). The barriers obtained with these metals were so low that they could only be deduced from J/V characteristics. The collected data from all four sets of experiments are shown in Fig. 2.19. Wherever possible, values of work functions obtained in ultra-high vacuum systems have been used. It is clear that the metals fall into two groups: those with work functions greater than about 4 eV, for which the barrier heights are all (with the exception of nickel) within 0.05 eV of 0.8 eV, and those with work functions less than about 4 eV, for which the barrier heights are all around 0.45 eV. Nickel has been found by many workers to give anomalously low barriers both on cleaved and etched silicon surfaces and on other semiconductors.

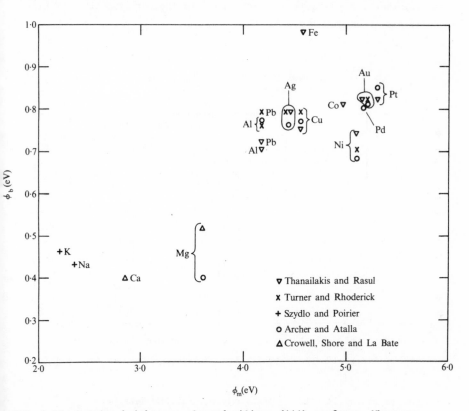

FIG. 2.19. Barrier heights on cleaved silicon (111) surfaces. Wherever possible, the work function values are those listed in Table 2.1 as having been obtained under u.h.v. conditions.

Some very careful measurements on cleaved silicon have recently been carried out by Thanailakis (1974, 1975) and Thanailakis and Rasul (1976). These authors cleaved silicon samples along (111) planes in a vacuum of 5×10^{-11} Torr, measured their work functions by means of the Kelvin vibrating-capacitor method, and then evaporated a metal film. The work function of the metal and the height of the resulting Schottky barrier were then measured without breaking the vacuum. The pressure during the evaporation was of the order of 10^{-10} Torr. The barrier height was measured from the J/V characteristics, by the photoelectric method, and from the C/V characteristics. The consistency of the measurements was very good; work functions could be determined to within ±0.05 eV and the barrier heights to within ±0.01 eV. The importance of this work is that, apart from the very high cleanliness and accuracy of the measurements, it avoids the difficulty inherent in the interpretation of all other data on cleaved surfaces, namely that of knowing the correct values of the work functions. Their results are given in Table 2.2.

TABLE 2.2.

Barrier heights of intimate contacts
(Thanailakis and Rasul)

| Metal | φ_m (eV) | φ_b (eV) | | $\Delta\varphi_b^*$ (eV) |
		from J/V characteristics and photoelectric measurements	from C/V characteristics	
Al	4.17	0.61	0.70	0.09
Fe	4.58	0.63	0.98	0.35
Co	4.97	0.61	0.81	0.20
Ni	5.10	0.59	0.74	0.15
Cu	4.55	0.62	0.75	0.13
Ag	4.41	0.68	0.79	0.11
Pt	5.30	0.71	0.82	0.11
Au	5.10	0.73	0.82	0.09
Pb	4.25	0.61	0.72	0.11

*$\Delta\varphi_b$ is the difference between φ_b from J/V and C/V characteristics.
The error for φ_m is approximately ± 0.05 eV and for φ_b is approximately ± 0.01 eV.

The following important points emerge from the data:
(a) In all cases there is excellent agreement between the values of φ_b obtained from the J/V and photoelectric measurements, but they are

both substantially smaller than the value obtained from C/V characteristics by an amount which is much greater than the image-force
lowering. It is not yet completely established whether this is an
artefact or an intrinsic property of an ideal metal—semiconductor
contact, but several indications suggest the latter. It will be
recalled from § 2.6 that the J/V and photoelectric methods both
measure the zero-bias barrier height φ_{b0}, while the C/V method
measures the flat-band barrier height φ_b^0. The image-force lowering
was negligible.

(b) The difference between the J/V and C/V values of the barrier height
is greatest for iron and decreases monotonically along the 3d
transition-metal series Fe—Co—Ni—Cu. Iron has the very large value
of 0.35 eV for $\Delta\varphi_b$.

(c) The barrier height is not a unique function of the metal work function. In particular, nickel and gold have the same work function
within experimental error, but their barrier heights differ by
about 0.1 eV, which is outside the experimental error.

(d) With the sequence Fe—Co—Ni, the barrier height decreases monotonically while the work function increases.

(e) The barrier heights obtained from the C/V data generally agree with
those of Archer and Atalla and of Turner and Rhoderick to within
about 0.05 eV. This is slightly worse than the combined estimated
errors of the respective sets of data (±0.03 eV). However, it must
be remembered that the vacuum used by Thanailakis and Rasul was much
better than that obtained by the other workers.

The effect of exposing a cleaved surface to oxygen before evaporating the
metal has been investigated by several workers. Archer and Atalla (1963)
studied the effect of cleaving silicon in oxygen at a pressure of 10^{-4} Torr
and exposing the surface to this ambient for a few seconds before evaporating the metal. They found a negligible effect on the barrier height for
platinum, palladium, nickel, and gold but a reduction of around 0.1 eV
for copper, silver, and aluminium. It is noteworthy that Turner and Rhoderick found no difference between the barrier heights on cleaved and etched
surfaces for gold and nickel (they did not investigate platinum and palladium) but a significant decrease on etched surfaces for copper, silver, and
aluminium. This suggests that exposure to oxygen has much the same effect
as etching. Crowell, Shore, and LaBate (1965) found that exposure of
cleaved silicon surfaces to oxygen, for a time which they estimated was
enough to produce about a monolayer coverage, lowered the barrier heights

of magnesium and calcium by about 0.08 eV. Mottram (1977) has observed the
effect of submonolayer contamination by oxygen using the same system as
Thanailakis and found a very slight reduction for gold. However, he found a
substantial reduction for silver and copper. He also found that the barrier
heights tended to move towards the values obtained on uncontaminated surfaces after about a week and that the difference observed by Thanailakis
between the values obtained from the J/V and C/V characteristics did not
exist with contaminated surfaces.

2.8.2. *Germanium*

There is very little information available about contacts to cleaved germanium surfaces, apart from some rather sparse results from Mead and Spitzer
(1964). They report barrier heights of 0.45 eV and 0.48 eV for gold and
aluminium, respectively, but the number of samples studied was extremely
small and they give very few details of their experimental procedure.

2.8.3. *III–V Compounds*

Again, because of the availability of good material, there is more information about contacts on cleaved surfaces of gallium arsenide than on any
other semiconductor apart from silicon. The first reported results were
those of Spitzer and Mead (1963), who made photoelectric and capacitance
measurements of the barrier heights of a series of metals evaporated on
cleaved (110) planes of both n- and p-type material. Similar measurements
were made by Smith (1969a). Both these sets of data are shown in Fig. 2.20.
Although the data are not as extensive as for silicon, there is a similarity
in that, for the lower work functions, barrier heights on cleaved surfaces

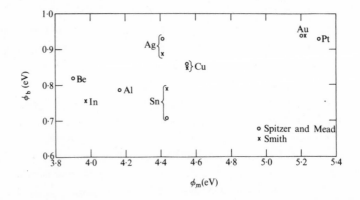

FIG. 2.20. Barrier heights on cleaved gallium-arsenide (110) surfaces.

tend to be higher than those on etched surfaces and there is less overall
variation from metal to metal.

Measurements on cleaved surfaces of other III—V compounds have been made
by Mead and Spitzer (1964), but in most cases they looked at only one or
two samples of each metal—semiconductor combination.

2.8.4. *Other semiconductors*

Spitzer and Mead (1963) have made an extensive series of measurements on
cleaved surfaces of cadmium sulphide. They find far more dependence of
barrier height on the choice of metal than exists with cleaved gallium
arsenide, although, when plotted against up-to-date values of work function,
they do not conform very well to the Mott limit. The same authors (Mead and
Spitzer 1964) have also made a preliminary survey of many other compounds,
but their data are rather sparse and not much detail is given. Their
results have been summarized by Mead (1966).

2.9. DISCUSSION

2.9.1. *General*

Two things stand out from the preceding considerations. First, the theory of
barrier heights is still in a rather rudimentary state, partly because of
its inherent difficulty and partly because of lack of knowledge of the precise physical nature of the interface. Second, the experimental measurements of barrier heights by different workers are not in general very consistent and, with the exception of silicon, tend to be rather sparse. There
is the further difficulty that, if the barrier height is considered as a
function of the metal work function, there is considerable doubt about what
value of the latter to use. This difficulty is not avoided, in the writer's
view, by using the electronegativity of the metal rather than its work
function. According to Pauling (1960), iron, cobalt and nickel all have the
same electronegativity, but they have different work functions and give
rise to different barrier heights when evaporated onto silicon. Bearing
these points in mind, it does not seem profitable to attempt to fit the
best straight line to the barrier-height versus work-function data by the
method of least squares, as some authors have done.

Nevertheless, certain generalizations can be made. Metals with high work
functions tend to produce larger barrier heights than those with low work
functions, although there is no smooth monotonic relationship between the
two. Also, for most semiconductors, the barrier heights are sensitive to

the method of preparation of the surface.

2.9.2. *Cleaved surfaces*

Contacts on cleaved surfaces generally show much less variation of barrier
height from metal to metal than contacts on etched surfaces covered by a
thin oxide layer, and a natural explanation of this can be given in terms
of surface states. Although the theory of intimate contacts is still in a
rather rudimentary stage, one might expect the barrier height on a cleaved
surface to be related to the density of intrinsic states on an ideal free
surface. For cleaved covalent semiconductors such as silicon, the total
number of surface states within the band gap is known to be about one per
surface atom (Many, Goldstein, and Grover 1965), so the density of states
would be expected to be of the order of 5×10^{18} eV^{-1} m^{-2}. This is consistent with the fact that the work function of cleaved silicon (111) surfaces is pinned by surface states, as discussed on p. 27, in such a way
that the Fermi level is about 0.3 eV above the valence band at the surface
(Allen and Gobeli 1962). When a semiconductor is in contact with a metal,
the surface states become the tails of the wave functions of the conduction electrons in the metal, as envisaged by Heine (1965) (see § 2.5), and
decay exponentially into the semiconductor with an attenuation length of
about 10 Å. In a very crude fashion one can adapt eqn (2.5) to the case of
intimate contacts by noting that the quantity $q\delta$ in the expression for γ
(eqn (2.6)) is the dipole moment of an interface state relative to the
surface of the metal and that, for an exponentially decaying charge density
the dipole moment is equal to q multiplied by the attenuation length. If,
in addition, one uses the permittivity ε_s of the semiconductor instead of
ε_i, one finds that γ should be about 0.1 for cleaved silicon surfaces. This
order of magnitude is compatible with those points in Fig. 2.19 which
correspond to work functions in excess of 4 eV.

However, one of the implications of Heine's theory is that the density
and distribution of surface states depend on the metal as well as on the
semiconductor.* This conclusion has been reinforced by Yndurain (1971), who
calculated a theoretical value of γ of about 0.2 for intimate contacts on
silicon. The experimental evidence concerning the effect of a metal on
surface states is inconclusive. Eastman and Freeouf (1975), using a photo-

---

*Surface states which are modified or induced by the presence of adsorbed
atoms or a layer of different material on the surface of a semiconductor
are called 'extrinsic' as opposed to the 'intrinsic' states on an ideal
free surface.

emission technique, observed surface states in the band gap of cleaved
(110) surfaces of gallium arsenide and other III—V compounds which remained
essentially unaltered after the deposition of two or three monolayers of a
metal. Moreover, they found in all cases a peak in the surface-state distribution within the band-gap, the low-energy edge of which lay just above
the position of the Fermi level in the corresponding metal—semiconductor
contacts. They conclude from this that, in contradiction to Heine and
Yndurain, the intrinsic surface states are unaffected by the metal and
determine the height of the Schottky barrier. However, Rowe, Christman, and
Margaritondo (1975), using an electron-energy-loss technique, found that
Eastman and Freeouf's result appeared to be a peculiarity of (110) surfaces
and that, on (111) surfaces of silicon, germanium and gallium arsenide, the
intrinsic states disappeared when a few monolayers of a metal were deposited. They explain the difference by supposing that the intrinsic surface
states on III—V compounds are predominantly associated with the group III
atoms, and that (110) surfaces become distorted in such a way that the
group III atoms are not in close contact with the metal.

If the experimental results for silicon are to be explained in terms of
surface states, whether intrinsic states or Heine tails which depend on the
presence of a metal, the data in Fig. 2.19 suggest that the energy distribution of these states has two peaks. As the metal work function decreases
from about 5.5 eV, the barrier height is first of all stabilized between
0.70 eV and 0.85 eV by a band of surface states located about 0.35 eV above
the valence band. When the work function has fallen to about 4 eV, this
band of states appears to be full, and their pinning action disappears. The
barrier height falls fairly rapidly as the work function falls from about
4 eV to 3.6 eV and is then apparently pinned at a value around 0.45 eV by a
second band of states located about 0.65 eV above the valence band, in
accordance with the mechanism discussed at the end of § 2.3.1. Intrinsic
surface states which are distributed between two bands have been calculated
theoretically and deduced experimentally by several authors for ideal silicon surfaces, though none of the reported distributions is of the form
described here. The theoretical origin of the two bands is connected with
the filled and unfilled bands discussed in § 2.1.3.

The results of Thanailakis (1975) and of Thanailakis and Rasul (1976)
are important because they establish beyond doubt that the barrier height
is not a smooth monotonic function of the work function of the metal but
depends on its detailed band structure. In particular, in the sequence
Fe—Co—Ni the barrier height actually decreases as the work function

increases; this cannot be explained in terms of surface states which are
an intrinsic property of the semiconductor alone. Louie and Cohen (1975)
have made a self-consistent pseudopotential calculation of the barrier
height at an ideal aluminium—silicon junction which, although it does not
incorporate an exact crystal potential in the aluminium, uses a 'jellium'
model with an electron density corresponding to that of aluminium. They
find states in the silicon energy gap which have the character of Bloch
waves in the aluminium and have exponentially decaying wave functions in
the silicon. They are similar to the states suggested by Heine, but retain
the characteristics of the intrinsic states which exist at the silicon
surface in the absence of the metal. They calculate the Al—Si barrier
height to be 0.6 ± 0.1 eV; this is in good agreement with Thanailakis's
values of 0.61 eV from J/V measurements and 0.70 eV from C/V data. They
have not yet extended their calculations to other metals. Tejedor, Flores,
and Louis (1977) have also made theoretical calculations of the Al—Si barrier height and find a value of 0.62 ± 0.05 eV.

It is not yet completely establishes whether the significant differences
between the barrier heights determined on the one hand from J/V and photoelectric measurements and on the other hand from C/V data, as reported by
Thanailakis and Rasul, are intrinsic properties of an ideal metal—
semiconductor contact or artefacts arising from the method of preparation.
The fact that they could be virtually eliminated by exposing the cleaved
surface to oxygen before evaporating the metal suggests the former.
Thanailakis and Rasul have attempted an explanation in terms of the
Heine model as portrayed in Fig. 2.12. The effect of the exponentially
decaying tails in the semiconductor is to distort the shape of the barrier
from its parabolic form as shown in Fig. 2.12(c). The J/V and photoelectric
methods measure the maximum height of the barrier φ_{b0} as seen by an electron, while the C/V method measures the extrapolated barrier height that
would correspond to the space charge of the ionized donors alone; i.e. as
if the barrier remained parabolic right up to the metal. This explanation
would imply some correlation between the difference in the barrier heights
$\Delta\varphi_b$ determined by the two methods and the density of states at the Fermi
surface of the metal, since the latter must affect the density of charge
in the tails. In this respect it may be significant that $\Delta\varphi_b$ decreases
monotonically in the sequence Fe—Co—Ni—Cu, which corresponds to the filling
of the 3d band. However, the present state of the theory is not sufficiently
advanced to put this explanation on a quantitative basis, and the question
remains open at the present time.

The barrier heights on cleaved gallium arsenide (Fig. 2.20) lie within
a remarkably narrow interval (between 0.75 and 0.95 eV) although the data
do not include such a wide range of metals as for silicon. Again the experimental data suggest pinning by surface states, but it is not so easy to
relate this to the density of intrinsic states on a free surface as it is
with silicon since, according to Van Laar and Sheer (1967), the work function of cleaved (110) surfaces of gallium arsenide is not pinned by surface
states, and theoretical calculations by Jones (1969) predict a gap in the
surface-state distribution. Mead and Spitzer (1964) found a similar narrow
range of barrier heights in other III—V compounds and formulated what is
sometimes called the 'one-third rule'; this states that, for group IV or
III—V semiconductors with the zincblende structure, the Fermi level in a
metal—semiconductor contact is located approximately $E_g/3$ above the
valence band. This is only a rough rule and there are quite a few III—V
compounds (e.g. indium phosphide, indium arsenide, and gallium antimonide)
which do not conform to it. There are insufficient data to say whether
germanium obeys the rule or not, and even silicon, which satisfies the rule
quite well as far as high-work-function metals are concerned, shows significant deviations for low-work-function metals such as sodium. It must be
remembered that the rule can only be expected to hold for cleaved surfaces.
It is essentially an empirical rule, though calculations by Pugh (1964)
of the surface-state distribution on diamond (111) surfaces have shown how
it might arise in the group IV semiconductors.

   Mead (1966) has suggested that semiconductors can be divided into two
categories — those in which the barrier height is pinned by surface states
and those in which surface states are unimportant. It is now clear that this
is too simple a view and that some semiconductors may belong to either
category according to the method of surface preparation. If only cleaved
surfaces are considered, Mead's classification has some rough validity. For
example, barrier heights on cleaved surfaces of cadmium sulphide are found
to show considerable dependence on the choice of metal (Spitzer and Mead
1963), whereas barriers on cadmium selenide do not (Mead 1965). Kurtin,
McGill, and Mead (1969) have extended this classification and suggest that
there is a gradual transition of a fundamental nature between covalent and
ionic crystals, one of the consequences of which is that surface states are
much more important on the former than on the latter. These authors have
compiled a graph showing the variation of a parameter S, which represents
the dependence of the barrier height on the electronegativity of the metal
for a particular semiconductor, with the difference in electronegativity $\Delta\vartheta$

of the constituent elements of the semiconductor. For elemental semiconductors, $\Delta\vartheta$ is zero and S is very small since the barrier height is insensitive to the choice of metal. For ionic semiconductors like zinc sulphide and zinc oxide, $\Delta\vartheta$ is equal to or greater than unity and S is close to unity, thus signifying that the barrier heights roughly conform to the Mott limit. The III–V compounds have small values of $\Delta\vartheta$ and S. This way of characterizing the properties of semiconductors is rather imprecise because the parameter S is not known or defined with any great accuracy, and the data appear to contain results for both cleaved and etched surfaces. Furthe more, some correction of the original data is necessary (Mead and McGill 1976). Nevertheless, it seems to be a valid way of describing an undoubted trend between the two extremes of covalent and ionic semiconductors. One can understand this in a qualitative way as arising because the outer-shell electrons are much more tightly bound in ionic than in covalent solids, so that the perturbing effect of the surface is much less pronounced.

2.9.3. Etched surfaces

A comparison of Fig. 2.15 and Fig. 2.19 shows that barriers on etched silicon surfaces are much more sensitive to the choice of metal than barriers on cleaved surfaces. The contrast is brought out more clearly in Fig. 2.16, which shows data from a single source. This difference can be easily understood in terms of surface states. It is well known from early studies of semiconductor surfaces that surface-state densities on 'real' (i.e. oxide-covered) surfaces are much lower than those on 'ideal' surfaces (see Many, Goldstein, and Grover 1965). This difference is often used in photoelectric studies of surface states to distinguish between the effects of surface and bulk states (Sebenne, Bolmont, Guichar, and Balkanski 1974), and can be understood in terms of dangling bonds by supposing that oxygen atoms make additional bonds with the semiconductor surface atoms so that all the orbitals are filled. A theoretical explanation using a band-structure approach has been given by Yndurain and Rubio (1971). With silicon, the difference in surface-state densities on the two types of surface may amount to two orders of magnitude or more.

From a theoretical point of view, one would expect contacts on etched surfaces to be easier to analyse because the oxide film to a certain extent 'decouples' the semiconductor from the metal, so that one can talk of interface states which are a property of the semiconductor–oxide combination. The band structure of the metal should be unimportant. If this is so, the Bardeen model should be reasonably good and the barrier height should

be a linear function of the metal work function as long as the surface-
state density is constant. As we have seen, the experimental data do not
really bear this out although a good deal of the scatter in the data must
be due to experimental uncertainties. If results from a single laboratory
using one particular method of surface preparation are considered (e.g.
Fig. 2.16), a straight line can often be roughly fitted to the data. For
etched silicon surfaces, the results of Turner and Rhoderick can be understood by postulating a surface-state density of about 10^{17} eV^{-1} m^{-2}, which
is nearly two orders of magnitude less than that required to explain their
results on cleaved surfaces but is greater than the values usually obtained
in MOS transistors. In the same way Smith's results on etched ($\bar{1}\bar{1}\bar{1}$) surfaces
of gallium arsenide can be reconciled with a surface-state density of about
4×10^{17} eV^{-1} m^{-2}.

Although the Bardeen model forms a useful basis for comparing theory
with practice, there are certain fundamental limitations to its validity.
The most questionable assumption is that the surface contributions to the
metal work function and semiconductor electron affinity are unchanged by
the presence of the interfacial layer (or, at any rate, that their difference is unchanged). A second simplifying assumption is that the surface
states can be represented by point charges, whereas in reality they must
have a spatial extent similar to that of the Heine tails. The model also
neglects any effect of image forces on the interface states. When it is
used to interpret experimental data, more practical limitations to its
validity arise from the assumptions that the properties of the interfacial
layer are the same for all samples, that there is no permanent charge in
the layer, and that the density of interface states is constant. Kar's
measurements (p. 56) are important in this context, first because his
values of $\varphi'_m - \chi'_s$ for silicon are not linearly related to the metal work
functions listed in Table 2.1, which shows that when the metal and the
silicon are separated by SiO$_2$ the surface contributions to φ_m are changed;
and secondly because the observed barrier heights for Mg and Sn, if inter-
preted in terms of the Bardeen model, can only be reconciled with Kar's
values of $\varphi'_m - \chi'_s$ by assuming an improbably high density of surface states.
These two considerations cast considerable doubt on the validity of the
model. Another serious criticism is that, as Archer and Yep have pointed
out, the Bardeen model predicts a much greater dependence of barrier height
on dopant concentration than is observed in practice.

A completely different model to explain the results on etched silicon
surfaces has recently been proposed by Andrews and Phillips (1975). These

authors use a chemical-bond approach and relate the barrier height to the heat of formation of the silicide which is formed when the metal forms chemical bonds with the silicon. They explain the fact that for gold contacts there is no difference between the barrier heights on cleaved and etched surfaces in terms of the extremely small heat of formation between gold and silicon. (In terms of the Bardeen model this can only be explained by postulating an accidental coincidence between the values of $\varphi_m - \chi_s$ and $E_g - \varphi_0$ for gold.) Their theory can be made to fit Turner and Rhoderick's data approximately if the dielectric constant of the oxide film is taken as ≈ 1.5, which is much lower than the value of four usually assumed for SiO_2.

3. Current-transport mechanisms

3.1. INTRODUCTION

This chapter discusses the transport mechanisms which determine the conduction properties of Schottky barriers. It assumes that a barrier has been
established as described in Chapter 2 and says nothing about the factors
which determine the height of this barrier except where they may affect the
J/V relationship.

The various ways in which electrons can be transported across a metal—
semiconductor junction under forward bias are shown schematically for an
n-type semiconductor in Fig. 3.1. The inverse processes occur under reverse
bias. The mechanisms are:

  (a) emission of electrons from the semiconductor over the top of the
      barrier into the metal,

  (b) quantum-mechanical tunnelling through the barrier,

  (c) recombination in the space-charge region,

  (d) recombination in the neutral region ('hole injection').

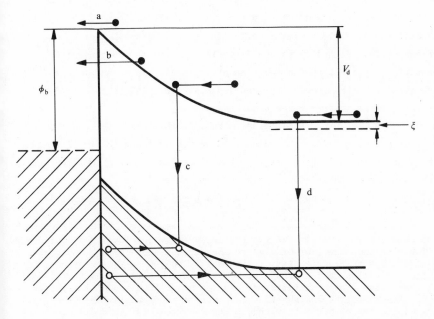

FIG. 3.1. Transport processes in a forward-biased Schottky barrier.

It is possible to make practical Schottky-barrier diodes in which (a) is the most important, and such diodes are generally referred to as 'nearly ideal'. Processes (b), (c), and (d) cause departures from this ideal behaviour.

3.2. EMISSION OVER THE BARRIER

3.2.1. *The two basic mechanisms*

Before they can be emitted over the barrier into the metal, electrons must first be transported from the interior of the semiconductor to the interface. In traversing the depletion region of the semiconductor, their motion is governed by the usual mechanisms of diffusion and drift in the electric field of the barrier. When they arrive at the interface, their emission into the metal is controlled by the number of Bloch states in the metal which can communicate with states in the semiconductor. These two processes are effectively in series, and the current is determined predominantly by whichever causes the larger impediment to the flow of electrons. According to the diffusion theory of Wagner (1931) and Schottky and Spenke (1939), the first of these processes is the limiting factor whereas, according to the thermionic-emission theory of Bethe (1942), the second is the more important.

The difference between the two theories is well brought out by the behaviour of the quasi-Fermi level for electrons in the conduction band of the semiconductor.* According to the diffusion theory, the concentration of conduction electrons in the semiconductor immediately adjacent to the interface is not altered by the applied bias. This is equivalent to assuming that at the interface the quasi-Fermi level in the semiconductor coincides with the Fermi level in the metal. In this case, the quasi-Fermi level droops down through the depletion region as shown in Fig. 3.2. This behaviour is in sharp contrast with the situation in a p–n junction under bias, where the quasi-Fermi levels for both types of carrier are generally assumed to be flat throughout the depletion region (Shockley 1950).

---

*The quasi-Fermi level ζ is a hypothetical energy level which is introduced to describe the behaviour of electrons under non-equilibrium conditions. It has the property that it predicts correctly the concentration of electrons in the conduction band if the electrons are assumed to be in thermal equilibrium at the lattice temperature and the quasi-Fermi level is used in the Fermi–Dirac distribution function in place of the equilibrium Fermi level. It also has the property that the electron current in the x direction is given by $qn\mu(d\zeta/dx)$, where n is the concentration of electrons and μ their mobility (see Shockley 1950).

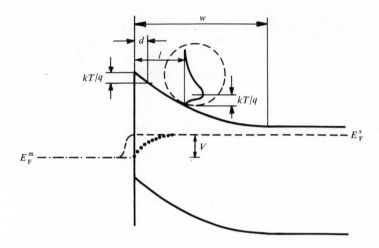

FIG. 3.2. Electron quasi-Fermi level in a forward-biased Schottky barrier
•••• according to diffusion theory, ----- according to thermionic-emission
theory. The broken circle shows the energy distribution of electrons which
make their last collision at a distance l from the interface.

The electrons emitted from the semiconductor into the metal are not in
thermal equilibrium with the conduction electrons in the metal but have an
energy which exceeds the metal Fermi energy by the barrier height (~ 1 eV).
They can be described loosely as 'hot' electrons, and the energy distribution of electrons on the metal side of the interface is as shown in Fig.
3.3. Gossick (1963) has suggested that the hot electrons in the metal can
be thought of as a species of electron different from the ordinary conduction electrons, and that they can be described by a quasi-Fermi level.

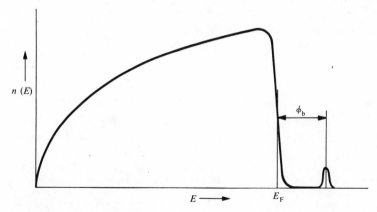

FIG. 3.3. Energy distribution of electrons emitted into the metal.

As the hot electrons penetrate into the metal, they lose energy by collisions with conduction electrons and the lattice and finally come into
equilibrium with the conduction electrons in the metal. Their quasi-Fermi
level falls until it ultimately coincides with the metal Fermi level.*
This process is rather like recombination in a semiconductor. This view-
point implies that the electron quasi-Fermi level at the interface does not
have to coincide with the metal Fermi level, and it is now possible to
envisage that the quasi-Fermi level remains flat through the depletion
region as in a p—n junction (see Fig. 3.2). This is the assumption made in
the thermionic-emission theory of Bethe (1942). Bearing in mind that the
gradient of the quasi-Fermi level provides the driving force for electrons,
one can see that, according to the diffusion theory, the main obstacle to
current flow is provided by the combined effects of drift and diffusion in
the depletion region whereas, according to the thermionic-emission theory,
the bottleneck lies in the process of emission of electrons into the metal.
In practice, the true behaviour will lie somewhere between the two extremes
of the diffusion theory and the thermionic-emission theory.

3.2.2. *The diffusion theory*

To derive the current/voltage characteristic according to the diffusion
theory, we write the current density in the depletion region in the usual
way as

$$J = qn\mu \mathcal{E} + qD_e \frac{dn}{dx}$$ (3.1)

where n is the concentration of electrons in the n-type semiconductor, μ
their mobility, D_e their diffusion constant, $\mathcal{E}$ the electric field in the
barrier, and $-q$ the charge on the electron.

It is only possible to write the expression for the current in the
simple form of eqn (3.1) if it is legitimate to use the concepts of a mobility and a diffusion constant which are independent of the electric
field. This is not the case near the top of the barrier where $\mathcal{E}$ has its
maximum value. Moreover, if the electron-distribution function changes
appreciably within the electron mean free path as is the case near the top

---

*Gossick's concept of a separate quasi-Fermi level for the hot electrons in
the metal is strictly valid only as long as they have an almost Maxwellian
velocity distribution. In practice this will not be so because they lose
their excess energy very rapidly through electron—electron collisions, and
it is perhaps more accurate to picture the electron quasi-Fermi level as
falling almost discontinuously to E_F^m at the metal—semiconductor boundary.

of the barrier, it is not even justifiable to split up the current into drift and diffusion components which are independent of each other. However, if such a simplification is not made, the analysis becomes extremely unwieldy; we shall therefore assume the validity of eqn (3.1), but it must always be borne in mind that the accuracy of the analysis ultimately rests on the truth of this assumption.

We now introduce the quasi-Fermi level for electrons ζ defined by

$$n = N_c \exp\{-q(E_c-\zeta)/kT\} \qquad (3.2)$$

where N_c is the effective density of states in the conduction band and E_c is the energy of the bottom of the conduction band. We have also used the Boltzmann approximation to the Fermi—Dirac function. Making use of Einstein's relationship $\mu/D_e = q/kT$, it is possible to write eqn (3.1) in the form

$$J = q\mu n \frac{d\zeta}{dx} \qquad (3.3)$$

which shows that the gradient of ζ supplies the 'driving force' for electrons. Combining eqns (3.2) and (3.3) gives

$$J = q\mu N_c \exp\{-q(E_c-\zeta)/kT\}\frac{d\zeta}{dx}$$

$$= kT\mu N_c \exp(-qE_c/kT)\frac{d}{dx}\{\exp(q\zeta/kT)\}. \qquad (3.4)$$

Integrating eqn (3.4) between $x = 0$ and $x = w$ gives

$$\frac{J}{kT\mu N_c} \int_0^w \exp(qE_c/kT)dx = [\exp(q\zeta/kT)]_0^w$$

$$= \exp\{q\zeta(w)/kT\}-\exp\{q\zeta(0)/kT\} \qquad (3.5)$$

Hence the current/voltage relationship is completely determined if E_c is known as a function of x and if the values of $\zeta(0)$ and $\zeta(w)$ can be specified for a particular value of applied bias. It is convenient to take the Fermi level in the metal as the zero energy level so that $\zeta(w) = V$, since the applied voltage V is equal to the difference between the Fermi levels at the terminals of the diode expressed in electron volts. The assumption

made in the diffusion theory is that the concentration of electrons on the
semiconductor side of the interface is unaltered by the application of bias;
i.e. that $\zeta(0) = 0$. Since the driving force for electrons is the gradient
of ζ, the assumption that $\zeta(0) = 0$ is equivalent to assuming that the impediment to current flow is provided entirely by the processes of drift and
diffusion in the depletion region.

   To evaluate the integral on the left-hand side of eqn (3.5), we shall
use the depletion approximation (see Appendix A) with a constant donor
density N_d. Eqn (A.5) then gives

$$E_c(x) = \varphi_b + \frac{qN_d}{2\varepsilon_s}(x^2 - 2wx) \tag{3.6}$$

where φ_b is the barrier height and ε_s ($=\varepsilon_{sr}\varepsilon_0$) is the permittivity of the
semiconductor. The integral may now be written as

$$\int_0^w \exp(qE_c/kT)\,dx = a^{-1}\exp(q\varphi_b/kT)\exp(-a^2w^2)\int_0^{aw}\exp(y^2)\,dy$$

$$= a^{-1}\exp(q\varphi_b/kT)F(aw) \tag{3.7}$$

where $a = (q^2N_d/2\varepsilon_s kT)^{\frac{1}{2}}$. $F(aw)$ is known as Dawson's integral and has been
tabulated numerically (e.g. Miller and Gordon 1931). If $aw > 2$, $F(aw)$ is
approximately equal to $(2aw)^{-1}$. (This approximation is equivalent to
neglecting the x^2 term in eqn (3.6) or assuming that $\&$ is constant and
equal to its maximum value $\&_{max}$ throughout the depletion region.) The condition $aw > 2$ is equivalent to $qV_d > 4kT$, and is generally satisfied except
for very large values of forward bias. Making this approximation, we can
now write eqn (3.5) as

$$J = 2kT\mu N_c a^2 w\{\exp(qV/kT)-1\}/\exp(q\varphi_b/kT)$$

$$= qN_c\mu\&_{max}\exp(-q\varphi_b/kT)\{\exp(qV/kT)-1\} \tag{3.8}$$

since, by Gauss's theorem, the maximum field strength is given by
$\&_{max} = qN_d w/\varepsilon_s = 2kTa^2w/q$. Eqn (3.8) gives the dependence of current on
voltage as predicted by the diffusion theory; it is almost, but not quite,
of the form of the ideal rectifier characteristic $J = J_0\{\exp(qV/kT)-1\}$.
The difference arises because $\&_{max}$ is not independent of bias voltage but
is proportional to $(V_{d0}-V)^{\frac{1}{2}}$. For large values of reverse bias, the current
does not saturate but increases roughly as $|V|^{\frac{1}{2}}$.

3.2.3. *The thermionic-emission theory*

In Bethe's thermionic-emission theory, the assumption is made that the current-limiting process is the actual transfer of electrons across the interface between the semiconductor and the metal. The inverse process under reverse bias is analogous to thermionic emission from a metal into a vacuum but with the barrier height φ_b replacing the metal work function φ_m. The effects of drift and diffusion in the depletion region are assumed to be negligible, which is equivalent to assuming an infinite mobility. It follows from eqn (3.3) that $d\zeta/dx$ is negligibly small so that the quasi-Fermi level for electrons remains flat throughout the depletion region and coincides with the Fermi level in the bulk semiconductor as in a p–n junction. This in turn implies that the concentration of electrons on the semiconductor side of the interface is increased by a factor $\exp(qV/kT)$ when a bias voltage V is applied. The situation can be visualized by imagining the existence of a membrane at the interface which the electrons can penetrate only with difficulty. This membrane keeps the electrons on the semiconductor side in thermal equilibrium with the bulk of the semiconductor. Inspection of Fig. 3.2 shows that the electron concentration on the semiconductor side of the boundary is now given by

$$n = N_c \exp\{-q(\varphi_b-V)/kT\}.$$

For a semiconductor with spherical constant-energy surfaces, these electrons have an isotropic distribution of velocities, and the number incident per second on unit area of the boundary is given by elementary kinetic theory (see, for example, Loeb 1961) as $n\bar{v}/4$, where $\bar{v}$ is the average thermal velocity of electrons in the semiconductor. The current density due to electrons passing from the semiconductor into the metal is therefore

$$J_{sm} = \frac{qN_c\bar{v}}{4}\{\exp -q(\varphi_b-V)/kT\}.$$

There is also a flow of electrons from the metal into the semiconductor which is unaffected by the application of bias because the barrier φ_b seen from the metal remains unchanged.* For zero bias, the current from semiconductor to metal just balances this current so that

---

*We are neglecting for the time being any possible field dependence of φ_b.

$$J_{ms} = \frac{qN_c \bar{v}}{4} \exp(-q\varphi_b/kT).$$

Hence

$$J = J_{sm} - J_{ms}$$

$$= \frac{qN_c \bar{v}}{4} \exp(-q\varphi_b/kT)\{\exp(qV/kT)-1\}. \tag{3.9}$$

For a Maxwellian distribution of velocities $\bar{v} = (8kT/\pi m^*)^{\frac{1}{2}}$, where m^* is the effective mass of electrons in the semiconductor. Substitution of $N_c = 2(2\pi m^* kT/h^2)^{3/2}$ then gives the current/voltage characteristic according to the thermionic-emission theory as

$$J = A^* T^2 \exp(-q\varphi_b/kT)\{\exp(qV/kT)-1\} \tag{3.10}$$

where

$$A^* = 4\pi m^* qk^2/h^3. \tag{3.11}$$

A^* is the same as the Richardson constant for thermionic emission except for the substitution of the semiconductor effective mass m^* for the free-electron mass m; it has the value

$$A^* = 1.2 \times 10^6 \ (m^*/m) \ \text{A m}^{-2} \ \text{K}^{-2} \tag{3.11a}$$

provided the conduction band of the semiconductor has spherical constant-energy surfaces. This is the case with gallium arsenide for which $(m^*/m) = 0.072$.

The approach we have adopted here follows the treatment given by Bethe in his original paper. There is an alternative approach to be found in the books by Henisch and Spenke according to which the electrons are assumed to have a Maxwellian distribution at the edge of the depletion region ($x = w$). The number of electrons which can surmount the barrier per second is then calculated on the assumption that the mean free path l is large compared with w. The expression for the current density calculated in this way is identical to eqn (3.9). It can easily be shown that the second approach is independent of the position of the particular reference plane

at which the electrons are assumed to have a Maxwellian distribution; in
other words, if the electrons were assumed to have a Maxwellian distribution at $x = x'$ ($x' \leqslant w$) which is in equilibrium with the bulk of the semiconductor, the current would be the same as if one took $x' = w$. If one
allows x' to approach zero (i.e. one assumes that the electrons are in
thermal equilibrium with the bulk of the semiconductor right up to the
metal, with a flat quasi-Fermi level), the two approaches are seen to be
equivalent. It follows that it is not necessary to assume that l is large
compared with w as is often done.

   The question of the correct value of m^* for a semiconductor with ellip-
soidal constant-energy surfaces is complicated and has been thoroughly discussed by Crowell (1965). He shows that the value of m^* associated with a
single energy minimum is

$$m^* = (l^2 m_y m_z + m^2 m_z m_x + n^2 m_x m_y)^{\frac{1}{2}},$$

where l, m and n are the direction cosines of the normal to the interface
relative to the principal axes of the ellipsoid and m_x, m_y, and m_z the
components of the effective-mass tensor. For semiconductors with spherical
constant-energy surfaces such as gallium arsenide, A^* is independent of
direction but, for semiconductors with anisotropic constant-energy surfaces
such as silicon and germanium, A^* depends on orientation even if the crystal
has cubic symmetry. For electrons in silicon, the appropriate values of
m^* obtained by summation over the six valleys are:

$$m^* = 2m_t + 4(m_l m_t)^{\frac{1}{2}} = 2.05m \qquad \text{for } \langle 100 \rangle \text{ directions}$$

and

$$m^* = 6\left(\frac{m_t^2 + 2m_l m_t}{3}\right)^{\frac{1}{2}} = 2.15m \qquad \text{for } \langle 111 \rangle \text{ directions}$$

where m_l and m_t are the longitudinal and transverse effective masses,
respectively. For electrons in germanium, the values are:

$$m^* = 4\left(\frac{m_t^2 + 2m_l m_t}{3}\right)^{\frac{1}{2}} = 1.19m \qquad \text{for } \langle 100 \rangle \text{ directions}$$

and

$$m^* = m_t + (m_t^2 + 8m_l m_t)^{\frac{1}{2}} = 1.07m \qquad \text{for } \langle 111 \rangle \text{ directions}.$$

The value of A^* for silicon is much larger than that for gallium arsenide, partly because of the larger effective-mass components and partly because of the six valleys in silicon.

An interesting and sometimes important feature of the thermionic-emission process is that the electrons injected into the metal under forward bias are confined to a narrow cone of directions. In the semiconductor, the electrons have an isotropic (or nearly isotropic if the constant-energy surfaces are not spherically symmetric) distribution of velocities with a mean velocity $\approx (3kT/m^*)^{\frac{1}{2}}$. When the electrons cross the interface, the component of velocity (strictly crystal momentum) parallel to the boundary is conserved, but the component perpendicular to the boundary is increased by an amount corresponding to the energy difference between the top of the barrier and the bottom of the conduction band in the metal; this is slightly greater than the Fermi velocity v_F. The electron trajectories in the metal are therefore confined to a cone or jet of half-angle $\tan^{-1}\{(3kT/m^*)^{\frac{1}{2}}/v_F\} \approx 5°$. Conversely, an electron may only pass from the metal into the semiconductor if its trajectory lies within this cone. For this effect to be present, one would expect the interface to have to be flat on a scale comparable to the wavelength of the electron wave function (~ 100 Å), since this is a necessary condition for conservation of the component of crystal momentum parallel to the boundary. This condition might be met in extremely high-quality interfaces (e.g. cleaved surfaces) but is unlikely to be satisfied by ordinary mechanically polished surfaces, so it is not clear how real the 'jet' effect is likely to be in practice.

3.2.4. The effect of the image force on the current/voltage relationship

The current/voltage relationship predicted by the thermionic-emission theory (eqn (3.10)) is of the form of the ideal rectifier characteristic $J = J_0\{\exp(qV/kT)-1\}$ with $J_0 = A^*T^2\exp(-q\varphi_b/kT)$, provided the barrier height is independent of bias. However, as we saw in Chapter 2, there are several reasons why the barrier height may depend on the electric field in the depletion region and hence on the applied bias. In particular, even in a perfect contact with no interfacial layer, the barrier height is reduced as a result of the image-force by an amount $\Delta\varphi_{bi}$ which depends on the bias voltage. The effective barrier that electrons must surmount before they can reach the metal can therefore be written as $\varphi_e = \varphi_b - \Delta\varphi_{bi}$. Moreover, as we saw in § 2.3.4, in the presence of an interfacial layer φ_b depends on the bias voltage so that φ_e may be bias dependent for two reasons. Such a bias dependence of φ_e will modify the current/voltage characteristic.

Let us suppose that $\partial\varphi_e/\partial V$ happens to be constant so that we may write $\varphi_e = \varphi_{b0} - (\Delta\varphi_{bi})_0 + \beta V$, where φ_{b0} and $(\Delta\varphi_{bi})_0$ refer to zero bias. The coefficient β is positive because φ_e always increases with increasing forward bias. The current density now becomes

$$J = A^*T^2\exp[-q\{\varphi_{b0}-(\Delta\varphi_{bi})_0+\beta V\}/kT]\{\exp(qV/kT)-1\}$$

$$= J_0\exp(-\beta qV/kT)\{\exp(qV/kT)-1\} \qquad (3.12)$$

where

$$J_0 = A^*T^2\exp[-q\{\varphi_{b0}-(\Delta\varphi_{bi})_0\}/kT].$$

We can write eqn (3.12) in the form

$$J = J_0\exp(qV/nkT)\{1-\exp(-qV/kT)\} \qquad (3.13)$$

where

$$\frac{1}{n} = 1 - \beta = 1 - (\partial\varphi_e/\partial V). \qquad (3.14)$$

n is often called the 'ideality factor'. If $\partial\varphi_e/\partial V$ is constant, n is also constant. For values of V greater than $3kT/q$, eqn (3.13) can be written as

$$J = J_0\exp(qV/nkT). \qquad (3.13a)$$

As we shall see, there are other mechanisms besides a bias dependence of φ_e which can give a current/voltage characteristic of this form. Eqn (3.13) is often written in the literature in the form

$$J = J_0\left\{\exp\left(\frac{qV}{nkT}\right)-1\right\}. \qquad (3.13b)$$

This form is incorrect because the barrier lowering affects the flow of electrons from metal to semiconductor as well as the flow from semiconductor to metal, so that the second term in the curly bracket must contain n. The difference between eqns (3.13) and (3.13b) is negligible for $V > 3kT/q$, but the correct form eqn (3.13) has the advantage that n can be found experimentally by plotting $\ln[J/\{1-\exp(-qV/kT)\}]$ against V. This graph should be a straight line of slope q/nkT if n is constant, even for

$V < 3kT/q$.

More usually $\partial\varphi_e/\partial V$ is not constant and the plot of $\ln[J/\{1-\exp(-qV/kT)\}]$ against V is not linear. The ideality factor defined by eqn (3.14) is now a function of bias, but it is still a useful concept which can be obtained from the experimental J/V characteristic through the relationship

$$\frac{1}{n} = \frac{kT}{q}\frac{d}{dV}\ln[J/\{1-\exp(-qV/kT)\}] \qquad (3.14a)$$

or, for $V > 3kT/q$,

$$\frac{1}{n} = \frac{kT}{q}\frac{d(\ln J)}{dV}. \qquad (3.14b)$$

The parameter n is generally a function of V, and can only be specified for a particular operating point on the characteristic. Eqn (3.14b) is more commonly found in the literature, but eqn (3.14a) has the advantage that it may be used for $V < 3kT/q$ and for reverse bias.

For the particular case where there is no bias dependence of φ_b, we have $(\partial\varphi_e/\partial V) = -(\partial\Delta\varphi_{bi}/\partial V)$ and $\Delta\varphi_{bi}$ is given by eqn (2.18a) as

$$\Delta\varphi_{bi} = \left\{\frac{q^3 N_d}{8\pi^2(\varepsilon_s')^2\varepsilon_s}\left(\varphi_b-V-\xi-\frac{kT}{q}\right)\right\}^{\frac{1}{4}}$$

so that, from eqn (3.14),

$$\frac{1}{n} = 1 - \frac{1}{4}\left(\frac{q^3 N_d}{8\pi^2(\varepsilon_s')^2\varepsilon_s}\right)^{\frac{1}{4}}\left(\varphi_b-V-\xi-\frac{kT}{q}\right)^{-\frac{3}{4}}.$$

If V is restricted to values less than $\varphi_b/4$, n is roughly constant. Taking $\varepsilon_s' = \varepsilon_s = 10^{-10}$ F m^{-1}, appropriate to silicon or gallium arsenide, and $\varphi_b - \xi \approx 0.5$ eV, n has a value of about 1.02 for $N_d = 10^{23}$ m^{-3}. The effect of image-force lowering on the forward characteristic is therefore negligible for Schottky barriers with $N_d < 10^{23}$ m^{-3}, although it may be more important under reverse bias because of the larger electric field in the depletion region.

3.2.5. *The combined thermionic-emission/diffusion theory*

Several authors (Schultz 1954, Gossick 1963, Crowell and Sze 1966) have combined the thermionic-emission and diffusion theories by considering

the two mechanisms to be in series and effectively finding the position of
the quasi-Fermi level at the interface which equalizes the current flowing
through each of them. The most fully developed theory is that of Crowell
and Sze, who introduce the concept of a 'recombination velocity' v_r at the
top of the barrier; v_r is defined by equating the net electron current into
the metal to $v_r(n-n_0)$, where n_0 is the equilibrium electron density at the
top of the barrier at zero bias.* The terms $v_r n$ and $v_r n_0$ represent the
electron flux from semiconductor to metal and the flux in the reverse
direction, respectively. The concept of a recombination velocity is quite
general and can be applied, for example, to a p–n junction. In terms of
the thermionic-emission theory outlined in § 3.2.3, which assumes that no
electrons which pass over the maximum of the potential barrier are scattered back into the semiconductor, $v_r = \bar{v}/4$ where $\bar{v}$ is the mean thermal
velocity of electrons in the semiconductor.

If we regard the position of the quasi-Fermi level $\zeta(0)$ at the interface as an adjustable parameter, the reasoning which leads up to eqn (3.9)
can be adapted to give the thermionic-emission current as

$$J_{te} = qN_c v_r \exp(-q\varphi_b/kT)\{\exp(q\zeta(0)/kT)-1\}. \tag{3.15}$$

Eqn (3.5) can be used to give the current limited by drift and diffusion in
the depletion region as

$$J_d = \frac{kT\mu N_c\{\exp(qV/kT)-\exp(q\zeta(0)/kT)\}}{\int_0^w \exp(qE_c/kT)\,dx} \tag{3.5a}$$

where we have used $\zeta(w) = V$. If the effect of the image force is taken
into account, it is the quasi-Fermi level $\zeta(x_m)$ at the maximum of the barrier which should be used rather than $\zeta(0)$, and the lower limit of inte-
gration should be x_m. Since the thermionic-emission current must equal the
current determined by drift and diffusion, $\zeta(0)$ can be eliminated between
eqns (3.15) and (3.5a) and, by writing $J_{te} = J_d = J$, we finally obtain

---

*If the barrier height is bias dependent, n_0 is also bias dependent and
is equal to the hypothetical electron density which would exist at the top
of the barrier if the barrier height at zero bias were equal to the barrier
height at bias V.

$$J = \frac{qN_c v_r \exp(-q\varphi_b/kT)\{\exp(qV/kT)-1\}}{1 + \dfrac{qv_r}{\mu kT} \exp(-q\varphi_b/kT) \displaystyle\int_0^W \exp(qE_c/kT)\,dx} \qquad (3.16)$$

which, following Crowell and Sze, may be written in the form

$$J = \frac{qN_c v_r}{1 + \dfrac{v_r}{v_d}} \exp(-q\varphi_b/kT)\{\exp(qV/kT)-1\} \qquad (3.17)$$

where

$$v_d = \left[\frac{q}{\mu kT} \exp(-q\varphi_b/kT) \int_0^W \exp(qE_c/kT)\,dx\right]^{-1}. \qquad (3.18)$$

v_d is the effective velocity due to drift and diffusion of electrons at the top of the barrier. The integral in the expression for v_d can be expressed in terms of Dawson's integral, as in eqn (3.7) and, if one adopts the same approximation as was used in deriving eqn (3.8), namely that the electric field is constant and equal to its maximum value $\mathcal{E}_{max}$, eqn (3.18) simplifies to $v_d = \mu\mathcal{E}_{max}$.

If $v_d \gg v_r$, eqn (3.17) reduces to eqn (3.9) with v_r in place of $\bar{v}/4$; the bottleneck to current flow is the actual emission of electrons into the metal, and the thermionic-emission theory applies. If $v_d \ll v_r$, eqn (3.17) reduces to eqn (3.8) when v_d is inserted as equal to $\mu\mathcal{E}_{max}$; the current is controlled by drift and diffusion in the depletion region and the diffusion theory applies. The condition for the validity of the thermionic-emission theory $(v_d \gg v_r)$ is equivalent to

$$\mu\mathcal{E}_{max} \gg \frac{\bar{v}}{4} \qquad (3.19)$$

i.e.

$$q\mathcal{E}_{max} \gg m^* \bar{v}/4\tau_c,$$

since $\mu = q\tau_c/m^*$, or

$$lq\mathcal{E}_{max} \gg kT, \qquad (3.20)$$

where τ_c is the mean time between collisions of electrons in the semiconductor and $l = \bar{v}\tau_c$ their mean free path. In deriving eqn (3.20), we have used $\bar{v} = 8kT/\pi m^*$ and have ignored a factor of $2/\pi$. Eqn (3.20) is equivalent to Bethe's criterion that the mean free path must exceed the distance d ($= kT/q\mathscr{E}_{max}$) in which the barrier decreases by an amount kT/q. The physical significance of this condition can be understood by reference to Fig. 3.2. Electrons which pass over the barrier into the metal will, on the average, have made their last collisions at a distance l from the maximum. Before making these collisions, the electrons will have a Boltzmann distribution of energies with an average energy of about kT (more accurately $\frac{1}{2}kT$) above the bottom of the conduction band as a result of motion normal to the interface. If $l \gg d$, only a small fraction of the electrons will have sufficient kinetic energy to negotiate the barrier, and those that do will be insufficient in number to change significantly the concentration of electrons at $x = l$. The electrons as a whole will therefore remain in equilibrium with the bulk of the semiconductor up to a distance l from the interface, as is assumed in the thermionic-emission theory. But, if $l \ll d$, nearly all the electrons will be able to negotiate the barrier, and the flow of electrons into the metal will result in a significant reduction in the concentration of electrons to a level below that of equilibrium with the bulk of the semiconductor; this is in accordance with the diffusion theory.

3.2.6. *Hot-electron effects*

At the top of the barrier the electric field may be quite large (typically $\sim 10^7$ V m^{-1}), and there are several hot-electron effects which impair the validity of the analyses in § 3.2.2 and § 3.2.5. First, we should not expect to be able to use values of mobility and diffusion constant which are independent of the electric field (Ryder and Shockley 1951). However, one cannot take the electric field into account simply by using a field-dependent mobility; with zero bias, the electrons are in thermal equilibrium with the lattice in spite of the built-in electric field and, as Stratton (1962) has shown, the effect of the field is to cool the electrons under forward bias when they are moving against the field and to heat them under reverse bias when they are moving with the field. Unfortunately, Stratton's analysis uses the boundary condition $\zeta(0) = 0$, which limits it to the case of the diffusion theory. The influence of hot-electron effects on the thermionic-emission theory has been discussed by Stokoe and Parrott (1974), who show that in a typical case the forward and reverse currents are both reduced by the order of 30%.

Secondly, since the energy-distribution function for electrons changes
significantly within an energy interval of kT, the condition $l \gg d$ implies
that the distribution function changes very considerably within a mean free
path, and in this case it is not possible to split up the current into drift
and diffusion components which are independent of each other as was assumed
in eqn (3.1). Hence the condition under which the thermionic-emission theory
should be valid ($l \gg d$) is precisely the condition under which eqn (3.1) is
not valid; this reinforces the comments made at the beginning of § 3.2.2.
This point has been discussed by Persky (1972) and Berz (1974) in the context of p—n junctions, and by Viktorovich and Kamarinos (1976) for Schottky
diodes.   But although it influences the condition for the validity of the
thermionic-emission theory, the exact mechanism of transport through the
depletion region does not affect the current/voltage characteristic in
those situations in which the thermionic-emission theory does apply.

A further weakness in Bethe's thermionic-emission theory is that it
assumes that the velocity distribution of electrons just inside the semiconductor at the top of the barrier is isotropic and Maxwellian, whereas in
fact it must be very anisotropic as electrons which enter the metal are
assumed not to return. (The hypothetical membrane postulated in § 3.2.3 is
in fact more like a sieve.) This point has been considered by Baccarani and
Mazzone (1976), who made a Monte-Carlo calculation of the velocity distribution. They find that the velocity distribution of electrons at the top of
the barrier is very close to a uni-directional Maxwellian distribution with
a mean velocity towards the metal which is twice that given by the Bethe
theory. However, the corresponding electron concentration is about half that
given by the Bethe theory, so that the resulting current density is within
about 1% of that given by eqn (3.10).

All these considerations show that we do not yet have an exact theory of
the current/voltage characteristic. A satisfactory treatment of electron
transport in a very rapidly varying potential is not yet available, and it
is surprising that the existing thermionic-emission theory works as well as
it does. However, the thermionic-emission theory is usually tested by comparing the value of the barrier height deduced from the J/V characteristic
with the results of other methods of measurement. Because a discrepancy of
a factor of two in A^* causes an error of less than kT/q in φ_b, the test is
not a very sensitive one.

3.2.7. *Refinements of the thermionic-emission theory*

Crowell and Sze (1966) incorporate several refinements into the theory, the

most important of which concern the value of the Richardson constant A^*.
A^* was first introduced in eqn (3.10) by using $v/4$ instead of Crowell and
Sze's recombination velocity v_r. This procedure assumes that all the
electrons incident on the interface at the potential maximum cross into
the metal and do not return. However, according to quantum mechanics, an
electron may be reflected by a potential barrier even if it has sufficient
energy to negotiate this barrier; furthermore, even after an electron has
passed over the barrier, it may be scattered with the absorption or emission of a phonon so that it returns from whence it came. The latter process
is most likely to take place between the metal and the position of the
potential-energy maximum as modified by the image force, i.e. within the
semiconductor. In this region, the bottom of the conduction band slopes
very steeply, and for most of its path within the semiconductor the
electron will have enough energy to emit an optical phonon. (The energy of
an optical phonon is about 0.06 eV in silicon and 0.04 eV in gallium arsenide.) For electrons with energies just above the threshold, the emission
of an optical phonon is more likely than any other form of scattering.
Crowell and Sze have calculated the probability f_p of an electron reaching
the metal without being scattered back into the semiconductor; the effect
of this on the thermionic-emission process is to multiply the recombination velocity v_r by a factor f_p, which is, of course, less than unity.
Their paper gives graphs of f_p as a function of the maximum electric field
$\mathscr{E}_{max}$ in the depletion region.* f_p approaches unity asymptotically as $\mathscr{E}_{max}$
increases because the position of the potential maximum gets nearer to the
metal, so that the distance within which phonon scattering is effective is
reduced; it decreases with increasing temperature because of the increasing phonon population. For silicon, f_p is about 0.95 at room temperature for
a field of 4×10^6 V m^{-1} which corresponds to a diffusion potential of
0.5 V and a donor density of 10^{22} m^{-3}. For gallium arsenide, the value is
somewhat lower, about 0.85, because of the larger optical phonon interaction
in polar compounds.

The effect of quantum-mechanical reflection has also been investigated
by Crowell and Sze, who included in their work the additional effect of
tunnelling of electrons through the top of the barrier. These two effects
combine to multiply the thermionic-emission current by a factor f_q which
depends on both temperature, through the temperature dependence of the

---

*The maximum electric field is the field due to the donors alone and
neglects the field due to the image.

electron energy, and on the maximum electric field $\mathcal{E}_{max}$ because this determines the height and shape of the potential barrier. There is some uncertainty in the calculations because an assumption has to be made about the
exact form of the image potential close to the surface of the metal. For
silicon and gallium arsenide, f_q rises above unity for fields in excess of
about 4×10^6 V m^{-1} because of the effect of tunnelling. Below this value,
f_q falls below unity because of reflection and approaches a value of about
0.6 for gallium arsenide and 0.5 for silicon.

Although Crowell and Sze's ideas about phonon scattering and quantum-
mechanical reflection are undoubtedly qualitatively correct, it is not
clear how reliable they may be quantitatively. The neglect of acoustic
phonon scattering must result in an over-estimate of f_p, and the use of the
effective-mass approximation in the calculation of f_q is of doubtful validity when the bottom of the conduction band slopes very steeply as it does
between the potential maximum and the surface of the metal. This latter
point has been considered by James (1949). For the case in which the
thermionic-emission theory is valid, the effect of f_p and f_q is to replace
A^* in eqn (3.10) by $A^{**} = f_p f_q A^*$. (In the general case the effect is to
replace v_r in eqn (3.17) by $f_p f_q v_r$.) Probably the most that can be said
with confidence is that A^{**} may be less than A^* by as much as 50%. In practice, because a variation in A^{**} by a factor of two has the same effect on
the current as a change in φ_b of less than kT/q, the difference between
A^* and A^{**} is not very important. Because f_p and f_q depend on the applied
bias (through $\mathcal{E}_{max}$), A^{**} is not strictly a constant, but this effect tends
to be masked in practice by image-force lowering of the barrier and other
effects. Crowell and Sze compute the effective values of A^{**} for electrons
in gallium arsenide and for silicon (111) surfaces at 300 K to be
4.4×10^4 A m^{-2} K^{-2} and 96×10^4 A m^{-2} K^{-2}, respectively; the large difference between these values is the result of the small effective mass in
gallium arsenide and the six equivalent valleys in the conduction band of
silicon. Andrews and Lepselter (1970) have re-examined Crowell and Sze's
computed values of f_p and f_q for silicon and find the best average value
of A^{**} to be slightly higher, namely 112×10^4 A m^{-2} K^{-2}. They also give
the value of 32×10^4 A m^{-2} K^{-2} for holes in p-type silicon.

3.2.8. *Comparison with experiment*

Crowell and Beguwala (1971) have calculated the position of the quasi-
Fermi level at the interface by eliminating J from eqns (3.15) and (3.5a),
and have concluded that, for semiconductors with a fairly high mobility

such as germanium, silicon, and gallium arsenide, the droop in the quasi-Fermi level through the depletion region $\{\zeta(w) - \zeta(0)\}$ is negligibly small, typically less than kT/q for a silicon diode with moderate forward bias. This justifies the fundamental assumption of the thermionic-emission theory. Their conclusion has been confirmed by Rhoderick (1972), who computed $\zeta(0)$ for several practical diodes by inserting the experimentally determined J/V characteristic into eqn (3.5a). There are therefore both theoretical and practical reasons for believing that Schottky diodes made from fairly high-mobility semiconductors should conform to the thermionic-emission theory rather than the diffusion theory.

Perhaps the most nearly ideal diodes yet reported are those of Arizumi and Hirose (1969), who made measurements on Schottky diodes made by evaporating Au, Ag, and Cu onto chemically cleaned silicon surfaces. Applying Rhoderick's analysis to the highest current densities reported in their paper, one finds that $\zeta(w) - \zeta(0) \approx 10^{-3}$ eV so that the diodes should conform to the thermionic-emission theory. The forward characteristics showed a linear dependence of $\ln J$ on V over more than four decades. n values were in the range 1.01–1.02, independent of temperature between 100 K and 350 K, and explicable in terms of image-force lowering of the barrier. The barrier heights obtained by assuming the thermionic-emission equation (3.10), with A^* replaced by A^{**} and given the value 96×10^4 Am^{-2} K^{-2}, agreed to within 2% with those obtained from C/V measurements and from the temperature dependence of the J/V characteristics. These data agree very closely with the predictions of the thermionic-emission theory. However, for large forward bias when the band bending is very small one would expect deviations from the pure thermionic-emission theory if v_d becomes comparable with v_r because of the reduction in $\mu\mathscr{E}_{max}$. Such behaviour is analogous to high-level injection in p–n junctions and has been observed by Wilkinson (1974) and by Wilkinson, Wilcock, and Brinson (1977) in titanium—silicon Schottky diodes.

Next to silicon, the semiconductor which has been most frequently used for Schottky-barrier studies is gallium arsenide, but here the data are much less conclusive. Kahng (1964) and Crowell, Sarace, and Sze (1965) have made gallium-arsenide Schottky diodes with $n \leqslant 1.04$. These values of n are greater than can be accounted for by image-force lowering alone, and the small difference may possibly be explained in terms of a component of current due to recombination in the depletion region (see § 3.4) or tunnelling through the top of the barrier (§ 3.3). Recombination is more important in gallium arsenide than in silicon because of the much shorter life-

time, and tunnelling is more important because of the very small effective
mass ($m^* = 0.07m$). Agreement between the barrier height determined from the
J/V characteristics by assuming Crowell and Sze's theoretical value of A^{**}
and the height deduced from the C/V measurements is not particularly good
(0.71 eV and 0.77 eV, respectively); the discrepancy is in the direction to
be expected if recombination or tunnelling were important and in the
opposite direction to what would be expected if the current were influenced
by the effect of diffusion in the depletion region. Although there are
still some unexplained discrepancies in the experimental data on gallium
arsenide diodes (e.g. Padovani and Sumner 1965), it does not appear that
these discrepancies can be resolved by diffusion effects, and Rhoderick's
analysis (1972) of data by Smith (1968a) shows that the controlling mechanism in moderately doped gallium-arsenide Schottky diodes at room temperature is almost certainly thermionic emission. This conclusion has recently
been confirmed by Gol'dberg, Posse, and Tsarenkov (1975), who found excellent agreement between the values of the barrier height of Au–GaAs diodes
as deduced from the J/V, C/V, and photoelectric characteristics. Their
experimental value of A^{**} was 8.6×10^4 A m^{-2} K^{-2} which agrees well with
the theoretical value given by eqn (3.11a) with $m^*/m = 0.072$, taking
$f_p f_q = 1$.

In conclusion, it appears that in Schottky diodes made from fairly high
mobility semiconductors the forward current is limited by thermionic emission provided the forward bias is not too large. Even comparatively low-
mobility semiconductors like GaP and CdS follow the thermionic-emission
theory quite closely (Rhoderick 1972), and eqn (3.19) seems to be unduly
severe as a condition for the validity of this theory. There is a dearth
of reliable data on low-mobility semiconductors which are likely to obey
the diffusion theory, and the only results known to the author which appear
to conform to the diffusion theory are some early data relating to CuO
(see Henisch 1957) and some recent measurements on amorphous silicon
(Wronski, Carlson, and Daniel 1976).

3.3. TUNNELLING THROUGH THE BARRIER

3.3.1. *Field and thermionic-field emission*

Under certain circumstances it may be possible for electrons with energies
below the top of the barrier to penetrate the barrier by quantum-
mechanical tunnelling. This may modify the ordinary thermionic process in
one of two ways which may be understood by reference to Fig. 3.4. In the

case of a very heavily doped (degenerate) semiconductor at low temperature, the current in the forward direction arises from the tunnelling of
electrons with energies close to the Fermi energy in the semiconductor.
This is known as 'field' emission. If the temperature is raised, electrons
are excited to higher energies and the tunnelling probability increases
very rapidly because the electrons 'see' a thinner and lower barrier. On
the other hand, the number of excited electrons decreases very rapidly
with increasing energy, and there will be a maximum contribution to the
current from electrons which have an energy E_m above the bottom of the
conduction band. This is known as 'thermionic-field' emission. If the
temperature is raised still further, a point is eventually reached in which
virtually all of the electrons have enough energy to go over the top of the
barrier; the effect of tunnelling is negligible and we have pure thermionic
emission.

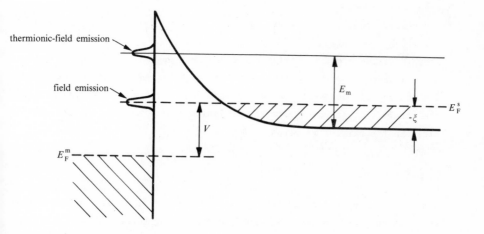

FIG. 3.4. Field and thermionic-field emission under forward bias. The diagram refers to a degenerately doped semiconductor for which ξ is negative.
(After Padovani and Stratton 1966.)

Tunnelling can be most easily understood if one recognizes the existence
of solutions of Schrödinger's equation which correspond to energies within
the forbidden gap of the semiconductor and which have the nature of exponentially damped waves rather than travelling waves. These can be most
simply visualized by supposing that the relationship $q(E-E_c) = \hbar^2 k^2/2m^*$,
which relates the kinetic energy of an electron to its wave vector k and
effective mass m^*, remains valid if the energy E of the electron is less

than the energy E_c of the bottom of the conduction band. In this case
the wave vector becomes imaginary and we may write $k = ik'$. Since the
electron wave function is proportional to $\exp(ikx) = \exp(-k'x)$, the wave
function decays exponentially with distance and the probability of finding
the electron at the position x, which is proportional to the square of the
wave function, is proportional to $\exp(-2k'x)$ where

$$\frac{\hbar^2 (k')^2}{2m^*} = q(E_c - E).$$

If $E_c - E$ is not constant (e.g. when the bands are not flat), k' is a
function of position and the probability of finding the electron at the
position x is given by $P = \exp(-2\int k' dx)$ and this is known as the WKB
approximation. If this result is used to calculate the probability of a
triangular barrier being penetrated by an electron with energy ΔE less than
the height of the barrier, it is found that

$$P = \exp\left\{-\frac{4}{3}(2qm^*)^{1/2}(\Delta E)^{3/2}/\hbar\mathscr{E}\right\} \tag{3.21}$$

where $\mathscr{E}$ is the electric field in the barrier. We can apply this to a
Schottky barrier with a diffusion potential V_d if ΔE is sufficiently small
for the top of the barrier to be thought of as triangular. Making use of
the result that, according to the depletion approximation, the maximum
field in the barrier is given by $\mathscr{E}_{max} = (2qN_d V_d/\varepsilon_s)^{\frac{1}{2}}$, we find that

$$P = \exp\left\{-\frac{2}{3}(\Delta E)^{3/2}/E_{00}V_d^{1/2}\right\} \tag{3.22}$$

where E_{00} (using the nomenclature of Padovani and Stratton 1966) is a parameter which plays an important role in tunnelling theory. It has the dimensions of energy divided by charge (or energy in eV) and is given by

$$E_{00} = \frac{\hbar}{2}\left[\frac{N_d}{m^* \varepsilon_s}\right]^{\frac{1}{2}} \tag{3.23}$$

$$= 18.5 \times 10^{-15}\left(\frac{N_d}{m_r \varepsilon_{sr}}\right)^{\frac{1}{2}} \text{ eV} \tag{3.23a}$$

where m^* $(= m_r m)$ is the effective mass of electrons in the semiconductor,
ε_s $(= \varepsilon_{sr}\varepsilon_0)$ its permittivity, and the donor concentration N_d is expressed
in m^{-3}. Values of E_{00} for semiconductors with various values of N_d, m_r,
and ε_r are shown in Fig. 3.5.

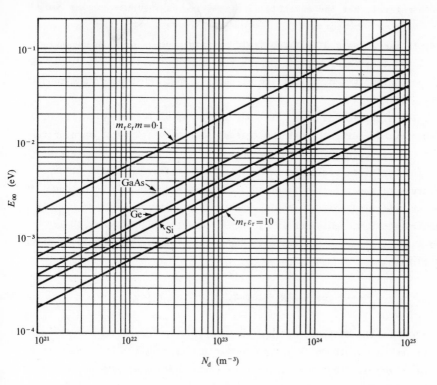

E_{00} (eV)

$m_r \varepsilon_r m = 0.1$

GaAs

Ge

Si

$m_r \varepsilon_r = 10$

N_d (m^{-3})

FIG. 3.5. Values of E_{00} as a function of N_d for various semiconductors.

For moderately doped semiconductors, the onset of thermionic-field emission can be considered to be roughly equivalent to a reduction in the barrier height by an amount $\Delta\varphi$ equal to $(\Delta E)_1$, where $(\Delta E)_1$ corresponds to a transmission probability of e^{-1}. The barrier lowering for which this occurs is

$$\Delta\varphi \approx \left(\frac{3}{2}\right)^{\frac{2}{3}} \left(E_{00}\right)^{\frac{2}{3}} \left(V_d\right)^{\frac{1}{3}} \qquad (3.24)$$

If we take as an example gallium arsenide with $N_d = 10^{22}$ m^{-3} and $E_{00} = 2 \times 10^{-3}$ eV and assume $V_d = 0.8$ V, we find that $\Delta\varphi \approx 0.02$ eV, which is barely observable. If, however, N_d is increased to 10^{23} m^{-3}, $\Delta\varphi \approx 0.04$ eV and becomes more significant.

The extension of the theory of tunnelling through Schottky barriers beyond the simple analysis given above is complicated and beyond the scope of this book. The theory has been developed by Padovani and Stratton (1966) and by Crowell and Rideout (1969); both these analyses are very mathematical, but the essential features are as follows:

(a) In the forward direction, field emission occurs only in degenerate semiconductors and, because of the very small effective mass, shows up at lower concentrations in gallium arsenide than in most other semiconductors.

(b) The significance of E_{00} is that it is the diffusion potential of a Schottky barrier such that the transmission probability of an electron with energy coinciding with the bottom of the conduction band at the edge of the depletion region is equal to e^{-1}. Therefore the ratio kT/qE_{00} is a measure of the relative importance of thermionic emission and tunnelling. As an approximation, we should expect field emission if $kT \ll qE_{00}$, thermionic-field emission if $kT \sim qE_{00}$, and thermionic emission if $kT \gg qE_{00}$. A more exact analysis shows that the temperature below which field emission occurs is given by

$$kT < 2qE_{00}\left\{\ln(-4\varphi_b/\xi)+(-2E_{00}/\xi)^{\frac{1}{2}}\right\}^{-1} \qquad (3.25a)$$

where ξ is the distance of the Fermi level below the bottom of the conduction band in the bulk semiconductor and is negative in a degenerate semiconductor. The temperature above which the process may be described as thermionic-field emission is given by

$$kT > 2qE_{00}\{\ln(-4\varphi_b/\xi)\}^{-1}. \qquad (3.25b)$$

The ranges of temperatures and concentrations over which gallium-arsenide Schottky barriers exhibit field and thermionic-field emission are shown in Fig. 3.6.

(c) The upper temperature limit for thermionic-field emission is given by

$$\cosh^2(qE_{00}/kT)/\sinh^3(qE_{00}/kT) < 2V_d/3E_{00} \qquad (3.26)$$

where V_d ($= \varphi_b - \xi - V$) is the diffusion potential. The transition between thermionic-field and pure thermionic emission therefore depends on the applied bias and is shown in Fig. 3.15 as a plot of diffusion potential against temperature.

(d) Except for very low values of forward bias, the current/voltage

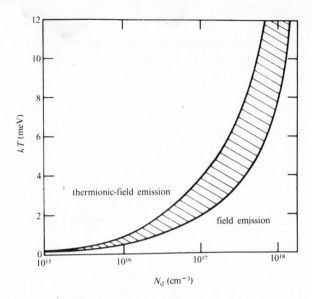

N_d (cm^{-3})

FIG. 3.6. Ranges of temperatures and concentrations over which gallium-arsenide Schottky barriers exhibit field and thermionic-field emission (Padovani 1971. Copyright Academic Press Inc.)

relationship is of the form

$$J = J_s \exp(V/E_0)$$ (3.27)

where

$$E_0 = E_{00} \coth(qE_{00}/kT).$$

Hence at low temperatures ($kT/qE_{00} \ll 1$), $E_0 \approx E_{00}$ so that the slope of the graph of ln J against V is independent of temperature. This is the case of field emission. At high temperatures ($kT/qE_{00} \gg 1$), E_0 is slightly greater than kT/q, and the slope of the graph of ln J against V can be written as q/nkT, where $n = qE_0/kT = (qE_{00}/kT)\coth(qE_{00}/kT)$. There is therefore a smooth transition from thermionic-field emission to pure thermionic emission.

(e) The pre-exponential term J_s is weakly dependent on the bias voltage. It is a complicated function of the temperature, barrier height, and semiconductor parameters for which the reader is referred to the original papers by Padovani and Stratton and by Crowell and Rideout.

(f) The maximum of the energy distribution of the emitted electrons occurs at an energy $E_m = V_d \{\cosh(qE_{00}/kT)\}^{-2}$ above the bottom of the conduction band in the bulk semiconductor.

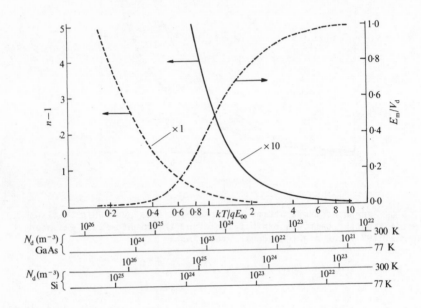

FIG. 3.7. Ideality factor n and position of maximum of energy distribution E_m of emitted electrons as a function of kT/qE_{00}.

Fig. 3.7 shows the variation of n and E_m/V_d as a function of kT/qE_{00} in the thermionic-field regime. If we somewhat arbitrarily take $n > 1.02$ and $E_m/V_d < 0.95$ as criteria for the transition from pure thermionic emission to thermionic-field emission, we see that it occurs when $kT/qE_{00} < 4$. In gallium arsenide this corresponds to $N_d > 10^{23}$ m^{-3} and 10^{22} m^{-3} at 300 K and 77 K, respectively, while in silicon the corresponding values of N_d are about a factor of four greater.

The analyses of Padovani and Stratton and of Crowell and Rideout both neglect the reduction of the barrier by the image force and the quantum-mechanical reflection of electrons with energies exceeding the height of the barrier. In a later paper Rideout and Crowell (1970) have taken both these effects into account for thermionic-field emission. They find a

significant increase in J_s and a small increase in n, primarily when the
conduction process is close to pure thermionic emission. Fig. 3.8 shows
their computed forward current/voltage characteristics presented in normalized form as a function of the band bending V_d (= $\varphi_b - V - \xi$).

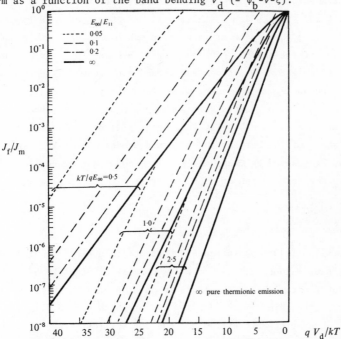

FIG. 3.8. Forward J/V characteristics for thermionic and thermionic-field
emission, allowing for image-force lowering and quantum mechanical reflection. The ordinate is the ratio of the current density to the 'flat-band'
current density $J_m = A^* T^2 \exp(-\xi/kT)$; the abscissa is the band bending
V_d (= $\varphi_b - V - \xi$) in units of kT/q. E_{11} is a measure of the modification of
the barrier by the image force and is equal to $(q^2/2\varepsilon_s)N_d^{\frac{1}{3}}$. E_{11} = 0 corresponds to no image-force lowering. (Rideout and Crowell 1970.)

Chang and Sze (1970) have taken both image-force lowering and quantum-
mechanical reflection into account and have also used Fermi–Dirac statistics so that their results are applicable to degenerate semiconductors.
However, they present their results in the form of curves computed especially for silicon, and their analysis cannot be readily adapted to other
semiconductors.

The general situation with regard to field and thermionic-field emission
is that the analyses of Padovani and Stratton and of Crowell and Rideout
seem to be adequate to explain the experimental data taking into account
the uncertainties arising from imperfect knowledge of the barrier para-

meters (see e.g. Padovani 1971). This is rather surprising in view of a
rather serious shortcoming of the theoretical model, namely that it ignores
the fact that the space charge associated with the donors is not continuous]
distributed but consists of discrete charges. For example, if one takes
$V_d \approx 0.5$ V and $N_d \approx 10^{23}$ m^{-3}, there will only be about four layers of donor
atoms (assuming them to be arranged in a cubic lattice) within the deple-
tion region. There are two effects. First, because the barrier consists of
a number of potential wells centred around each donor, its shape will no
longer be parabolic so the tunnelling probability will be altered; secondly,
because the number of donors within the deplation layer is small and they
are not arranged in a cubic lattice, there will be fluctuations in the
precise shape of the barrier and an average has to be taken. The second
point was considered by Chang and Sze who found that it could lead to
errors in the calculation of J of the order of 10% for $N_d = 10^{24}$ m^{-3} and
100% for $N_d = 10^{25}$ m^{-3}.

3.3.2. *Ohmic contacts*

Field emission is of considerable importance in connection with ohmic contacts to semiconductors, which usually consist of Schottky barriers on very
highly doped material. What matters in this case is the differential resistance around zero bias (i.e. $(\mathrm{d}V/\mathrm{d}J)_{V=0}$) for which eqn (3.27) is not valid.
The full theoretical expression for the contact resistance R_c is complicated
and has been discussed by Yu (1970). He shows that the dependence of R_c on
doping level is determined predominantly by the factors

$\exp(\varphi_b/E_{00})$ for field emission $(qE_{00}/kT \gg 1)$

$\exp\{\varphi_b/E_{00} \coth(qE_{00}/kT)\}$ for thermionic-field emission $(qE_{00}/kT \sim 1)$

$\exp(q\varphi_b/kT)$ for thermionic emission $(qE_{00}/kT \ll 1)$.

Thus for low doping (the thermionic-emission regime) R_c is independent of
N_d, while for highly doped material the conduction process is field emission and $\ln R_c$ is proportional to $N_d^{-\frac{1}{2}}$. The thermionic-field regime bridges
the two. Fig. 3.9 shows some experimental determinations made by Hooper,
Cunningham, and Harper (1965) of R_c plotted as a function of N_d for n-
type silicon at room temperature. The theoretical curves are those computed by Vilms and Wandinger (1969) by using either a simple parabolic
barrier, the height of which was adjusted to allow for the image force

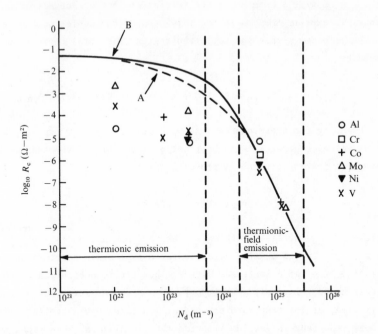

FIG. 3.9. Contact resistance against doping density for contacts to n-type silicon (Vilms and Wandinger 1969). Experimental data by Hooper, Cunningham, and Harper (1965). Curves A and B refer to two different theoretical models (see text).

(curve A), or a truncated parabolic barrier (curve B). The agreement between theory and experiment is reasonable.

3.4. RECOMBINATION IN THE DEPLETION REGION

The importance of recombination in the space-charge region has been convincingly demonstrated in a classic paper by Yu and Snow (1968). The recombination normally takesplace via localized centres and, according to theory (Shockley and Read 1952, Hall 1952), the most effective centres are those with energies lying near to the centre of the gap. The theory of the current due to such a recombination centre is the same for a Schottky diode as for a p—n junction (Sah, Noyce, and Shockley 1957), and the current density is given approximately by

$$J_r = J_{r0}\{\exp(qV/2kT)-1\} \tag{3.28}$$

where $J_{r0} = qn_i w/2\tau_r$. Here n_i is the intrinsic electron concentration proportional to $\exp(-qE_g/2kT)$, w is the thickness of the depletion region, and
τ_r is the lifetime within the depletion region. The total current density
is given by

$$J = J_{te} + J_r = J_{t0}\{\exp(qV/kT)-1\} + J_{r0}\{\exp(qV/2kT)-1\}$$

where, assuming the thermionic-emission theory, $J_{t0} = A^{**}T^2\exp(-q\varphi_b/kT)$.
For bias voltages greater than about $4kT/q$, the ratio of the thermionic
to the recombination current is proportional to

$$\tau_r\exp\{q(E_g+V-2\varphi_b)/2kT\}.$$

The ratio J_{te}/J_r increases with τ_r, V, and E_g, and decreases with φ_b. Also,
since $E_g + V - 2\varphi_b$ is usually negative for n-type semiconductors and small
values of V, the ratio increases with T. Thus the recombination component
is likely to be relatively more important in high barriers, in material of
low lifetime, at low temperatures, and at low forward-bias voltage. It is
much more important in gallium arsenide than in silicon. When recombination
current is important, the temperature variation of the forward current
shows two activation energies* [fig. 3.10]. At high temperatures the activation energy tends to the value φ_b - V characteristic of the thermionic-
emission component; at low temperatures it approaches the value $(E_g-V)/2$
characteristic of the recombination component.

Recombination current is a common cause of departure from ideal behaviour in Schottky diodes. Fig. 3.11 shows a hypothetical diode characteristic which has been synthesized by adding a recombination term

$$I_r = 3 \times 10^{-9}\{\exp(x/2)-1\} \text{ A}$$

to the ideal thermionic-emission term

$$I_{te} = 10^{-9} \times \{\exp(x)-1\} \text{ A}.$$

Here $x = q(V-20I)/kT$ corresponding to a series resistance of 20 Ω. The
resultant characteristic is well fitted over four decades by

---

*Any process which shows a temperature dependence of the form $\exp(-E_a/kT)$
is said to have an activation energy E_a.

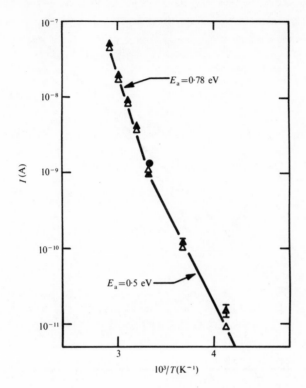

FIG. 3.10. Forward current of platinum–silicon Schottky diode as a function of temperature (Yu and Snow 1968).

$$I = 2 \times 10^{-9}\{\exp(qV/1.07kT)-1\} \text{ A.}$$

Recombination current may therefore cause deviations of n from unity and of the pre-exponential term from the ideal value $A^{**}T^2\exp(-\varphi_b/kT)$. The importance of recombination current in causing small departures from ideal behaviour has been frequently overlooked in the literature. Such departures become more pronounced at low temperatures.

3.5. HOLE INJECTION

If the height of a Schottky barrier on n-type material is greater than half the band gap as is often the case, the region of the semiconductor adjacent to the metal becomes p-type and contains a high density of holes. One might expect some of these holes to diffuse into the neutral region of

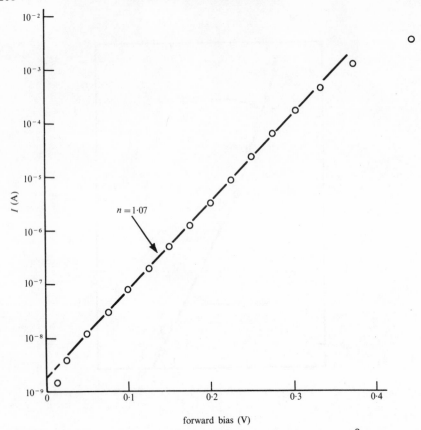

FIG. 3.11. The circles are plotted from the function $I = 10^{-9}\{\exp(x)-1\} + 3 \times 10^{-9}\{\exp(x/2)-1\}$ where $x = q(V-20I)/kT$. The solid line represents the function $I = 2 \times 10^{-9}[\exp(qV/1.07kT)-1]$.

the semiconductor under forward bias, thus giving rise to the injection of holes. Hole injection at metal contacts was extensively studied in the early days of semiconductors, and the early work has been summarized by Henisch (1957). It is convenient to distinguish between the behaviour of plane contacts formed by evaporation and that of point contacts.

3.5.1. Hole injection in plane contacts

If one assumes that the hole quasi-Fermi level is flat through the depletion region and coincides with the Fermi level in the metal, the hole current density can be written, using ordinary p–n junction theory, as

$$J_h = \frac{qD_h p_0}{L}\{\exp(qV/kT)-1\} \tag{3.29}$$

where p_0 $(= n_i^2/N_d)$ is the equilibrium concentration of holes at the edge of
the depletion region, and D_h and L are the diffusion constant of holes in
the bulk semiconductor and the thickness of the quasi-neutral region,
respectively. (L is assumed to be less than the hole diffusion-length.)
The assumption of a flat quasi-Fermi level has been justified by Rhoderick
(1972). Since the electron current is given by eqn (3.10) if one assumes
the thermionic-emission theory, the hole injection-ratio γ_h is given by

$$\gamma_h = \frac{J_h}{J_h + J_e} \approx \frac{J_h}{J_e} = \frac{qD_h n_i^2}{N_d LA^{**} T^2 \exp(-q\varphi_b/kT)} \tag{3.30}$$

a result which seems to have been first derived by Henisch. The injection
ratio increases with φ_b because of the reduction in J_e, and decreases
with N_d because of the decrease in p_0 and consequent reduction in J_h. If
one substitutes typical figures of $\varphi_b \approx 0.8$ eV, $N_d \approx 10^{22}$ m^{-3}, and
$L \approx 5 \times 10^{-6}$ m for silicon diodes, one finds $\gamma \approx 10^{-4}$ so that hole injection is generally negligible. It is for this reason that Schottky diodes
are often described as 'majority-carrier' devices.

Eqn (3.29) takes into account only the diffusion component of the hole
current density. Scharfetter (1965) has pointed out that, for large forward bias, the electric field in the quasi-neutral region of the semiconductor gives rise to a significant drift component, and for very large
currents the injection ratio rises linearly with J. The critical current
density at which this increase in γ_h begins to take place is given by
$J_c = qD_e N_d/L$, where J_c is typically of the order 10^5 A m^{-2} in practical
diodes. Scharfetter's theory has recently been extended by Green and
Shewchun (1973) who have found that there is a limit to the growth of γ_h
with J. This limit arises partly because of high-level-injection effects
when the minority-carrier density in the neutral region becomes comparable
with the majority-carrier density, and partly because the supply of holes
at the contact is limited by thermionic emission. The result of these two
effects is that the injection ratio may pass through a maximum and decrease
at high current densities.

Scharfetter's theory has been partially confirmed by some experimental
results of Yu and Snow (1969) on gold–silicon diodes. They found that at
low current densities the experimental value of the injection ratio agreed
well with eqn (3.30) and that γ_h began to increase when J approached the
value J_c given above, as shown in Fig. 3.12. However, the measurements were

not continued to sufficiently high current densities to establish whether γ_h increases linearly with J in agreement with Scharfetter's theory, or passes through a maximum as predicted by Green and Shewchun.

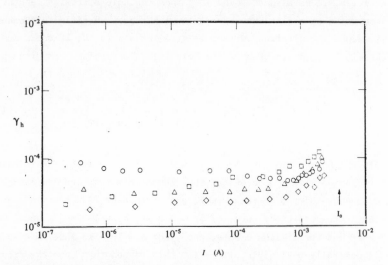

FIG. 3.12. Injection ratio γ_h ($= J_h/J$) as a function of current for a gold–silicon Schottky diode (Yu and Snow 1969). Emitter area = 3.075 × 10^{-3} cm^2 and $N_d = 10^{16}$ cm^{-3}.

If hole injection becomes appreciable at high current densities, one might expect that the additional electrons which must enter the neutral region to maintain charge neutrality would increase the conductivity and reduce the series resistance. Conductivity modulation arising from this cause has been reported by Jäger and Kosak (1973) as occurring in high-power Schottky rectifiers; it has also been reported by Andersson, Hyder, and Berg (1973) in silicon particle detectors. The effect is noticeable only with large barrier heights and in weakly doped material. Hole injection is also important when Schottky diodes are used under 'punch-through' conditions as in BARRITT diodes (Sze, Coleman, and Loya 1971). A further discussion of hole injection is given in Appendix C.

3.5.2. Hole injection in point contacts

The minority-carrier injection ratio in a point-contact rectifier may be quite large (approaching unity), and the operation of point-contact transistors depended on this fact. The reason for this difference in behaviour is not clear. All the work on point-contact transistors was carried out

using germanium, and there is almost no reliable information available
about the relevant barrier heights; it is therefore impossible to say
whether eqn (3.30) would predict a large or a small injection ratio. Eqn
(3.30) is in fact quite sensitive to φ_b, N_d, and E_g (the latter through
n_i^2). However, the conditions of operation of point-contact rectifiers are
vastly different from those pertaining to plane contacts especially with
regard to the current density. In their original paper on the point-
contact transistor, Bardeen and Brattain (1949) quote currents of $\sim 10^{-3}$ A
flowing in tips of radius 5 μm; this corresponds to a current density of
about 10^7 A m^{-2}. With such a high current density the change in potential
at the edge of the depletion region due to the forward bias must be quite
large and the bands almost flat so that eqn (3.30) does not apply. Under
such conditions the hole current is determined more by the transport processes in the quasi-neutral region than by the contact itself, and values
of γ_h approaching unity are possible as was shown by Braun and Henisch
(1966). According to Clarke, Green, and Shewchun (1974), the hole current
is accentuated by the spherical symmetry causing the hole concentration at
a distance r to decrease as r^{-1} and so increasing the diffusion rate.

3.6. REVERSE CHARACTERISTICS

According to the thermionic-emission theory, the reverse current density of
an ideal Schottky diode should saturate at the value $J_0 = A^{**} T^2 \exp(-q\varphi_b/kT)$.
There are several causes of departure from this ideal behaviour, and these
are outlined in the following sections.

3.6.1. *Field dependence of the barrier height*

If for any reason φ_b is dependent on the electric-field strength in the
barrier, the reverse characteristics will not show saturation. There are
several possible mechanisms, all of which predict that φ_b should be a
decreasing function of $\mathscr{E}_{max}$, the maximum field strength in the barrier.
Since $\mathscr{E}_{max}$ increases with reverse bias V_r, it follows that φ_b decreases
with increasing V_r, and the current does not saturate but increases pro-
portionally to $\exp(q\Delta\varphi_b/kT)$, where $\Delta\varphi_b$ is the lowering of the barrier due
to the field.

The simplest form of barrier lowering is that due to the image force.
The corresponding value of $\Delta\varphi_b$ is given by eqn (2.18a) with V replaced by
$-V_r$. Since $\Delta\varphi_{bi}$ is proportional to $V_r^{\frac{1}{4}}$ for large values of reverse bias, a
plot of ln J against $V_r^{\frac{1}{4}}$ should give a straight line, the intercept of which
on the ln J axis gives J_0. Arizumi and Hirose (1969) have described silicon

Schottky diodes with a donor density of 10^{21} m^{-3} in which the reverse characteristics can be completely explained in terms of image-force lowering up to a reverse bias of 10 V, and these diodes appear to be the most nearly ideal that have yet been made.

More usually the barrier lowering necessary to explain the lack of saturation is considerably greater than that due to the image force. One of the commonest causes of a field-dependent barrier height is the presence of an interfacial layer, and this will be discussed in § 3.8.

Andrews and Lepselter (1970) observed a lack of saturation in the reverse characteristics of silicon—silicide Schottky diodes which could be explained in terms of a field-dependent barrier height. These diodes were made by depositing Rh, Zr, or Pt on silicon and heating the combination to such a temperature that the metal reacted with the silicon to form a silicide. All these silicides exhibit metallic conductivity, and a Schottky barrier is formed between the silicon and the metal silicide which is thought to be free from any interfacial layer. Andrews and Lepselter were able to explain their reverse characteristics by assuming an empirical field dependence of the form $\Delta\varphi_b = \alpha'\mathscr{E}_{max}$ with values of α' in the range 15—35 Å. An example of how successfully this empirical analysis fits their data is given in Fig. 3.13. The origin of the term α' was presumed not to be the same as that of the parameter α introduced in § 2.3.2 because of the absence of an interfacial layer.

A possible explanation of the $\alpha'\mathscr{E}_{max}$ term has been postulated by Andrews and Lepselter and others (Parker, McGill, Mead, and Hoffman 1968; Broom 1971) as follows. The silicide is supposed to form a perfect metal—semiconductor junction with the silicon and, according to the Heine model (§ 2.5), the wave functions of the conduction electrons in the metal penetrate into the forbidden gap of the semiconductor in the form of exponentially damped evanescent waves. These exponential tails constitute an electric dipole which distorts the barrier shape in such a way that its height is reduced, and to a first approximation the reduction is proportional to $\mathscr{E}_{max}$. But although this explanation may be qualitatively correct, there are difficulties in applying it in a quantitative manner. According to both Parker *et al.* and Broom, the position of the maximum of the potential barrier required to explain their data is about 50 Å from the interface. It is a very poor approximation to assume that the tails of the conduction-electron wave functions can be represented by a simple exponential over this whole distance when the bands are bent as drastically as they are near the potential maximum; it is also not clear how the concept of a

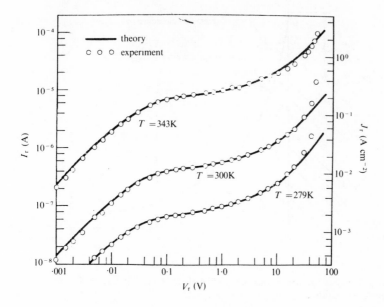

FIG. 3.13. Reverse characteristics of a ZrSi$_2$–Si Schottky diode. The theoretical curve is based on a field-dependent barrier height of the form $\varphi_b = \varphi_b^0 - \alpha'\mathscr{E}_{max}$, with $\alpha' = 15$ Å (Andrews and Lepselter 1970).

'neutral level' (§ 2.1.3) comes into this picture. Furthermore, both Andrews and Lepselter and Parker *et al.* assume that the contributions due to the image force and to wave-function penetration are additive. This is only true if the effect of the image force is negligibly small; in general, the two effects interfere with each other and are not additive. A completely satisfactory analysis of the effect of wave-function penetration has yet to be given.

3.6.2. *The effect of tunnelling*

Tunnelling through the barrier becomes significant at lower doping levels in the reverse direction than in the forward direction because the bias voltages involved are much greater; also the application of a moderately large reverse bias can cause the potential barrier to become thin enough for significant tunnelling of electrons from the metal to the semiconductor to take place. The effect of tunnelling can again be described as either field or thermionic-field emission as shown in Fig. 3.14.

Fig. 3.15 shows the boundaries between the thermionic and thermionic-

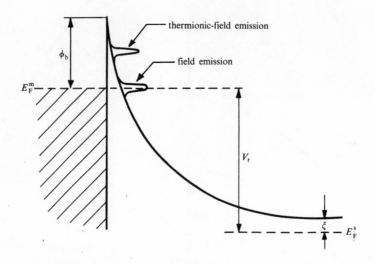

FIG. 3.14. Field and thermionic-field emission under reverse bias (non-degenerate case).

field regimes on a plot of band bending V_d against temperature with E_{00} as a parameter (Padovani and Stratton 1966). The curves are based on eqn (3.26), and apply to both forward and reverse bias. It can be seen from this Figure that, at room temperature for reverse biases around 3 V, departures from pure thermionic emission begin at donor concentrations greater than 10^{23} m^{-3} for silicon (corresponding to E_{00} = 3 meV), compared with 5×10^{23} m^{-3} for forward bias. In the thermionic-field regime the current/voltage relationship has been derived by Padovani and Stratton as

$$J = J_s \exp(V_r/E') \tag{3.31}$$

where

$$E' = E_{00}\{(qE_{00}/kT) - \tanh(qE_{00}/kT)\}^{-1}$$

and

$$J_s = \frac{A^* T (\pi q E_{00})^{\frac{1}{2}}}{k} \left\{ q(V_r - \xi) + \frac{q\varphi_b}{\cosh^2(qE_{00}/kT)} \right\}^{\frac{1}{2}} \exp(-\varphi_b/E_0).$$

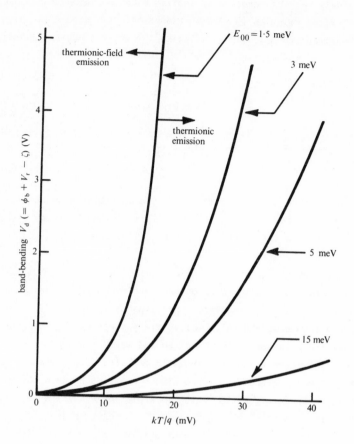

FIG. 3.15. Boundaries between thermionic and thermionic-field regimes on a plot of band bending V_d against temperature, with E_{00} as a parameter. The curves are derived from Padovani and Stratton's eqn (51) modified as follows: $-E$ is replaced by $-E + \xi_2$ (their notation, see Padovani 1971, p. 111), and E_B is replaced by $-E_B$. The latter modification seems to be necessary because the condition for thermionic-field emission must depend on the band bending V_d, which is equal to $E_B + \xi_2 - E$ in their notation, since V_d determines the shape of the potential barrier. If these modifications are made, their eqn (51) is identical to their eqn (33) and their Fig. 13 becomes identical with their Fig. 3 when expressed in terms of V_d. The two figures must coincide for small values of bias in either direction. The curves drawn here therefore apply to both forward and reverse biases, values of V_d less than $\varphi_b - \xi$ (our notation) denoting forward bias (i.e. negative V_r).

J_s is a slowly varying function of applied bias. A simple analysis of thermionic-field emission in reverse-biased Schottky diodes has also been made by Honeisen and Mead (1971) in which the current density is expressed as a function of the maximum electric-field strength $\mathscr{E}_{max}$. It is possible to express $\mathscr{E}_{max}$ in terms of E_{00} and the band bending $\varphi_b + V_r - \xi$, in which case Honeisen and Mead's expression becomes

$$J = \frac{2A^*(\pi T)^{\frac{1}{2}}\{q(\varphi_b+V_r-\xi)\}^{\frac{1}{2}}qE_{00}}{k^{3/2}} \exp\left\{-\frac{q\varphi_b}{kT} + \frac{q^3(\varphi_b+V_r-\xi)E_{00}^2}{3(kT)^3}\right\} \tag{3.32}$$

which, for small values of qE_{00}/kT, shows the same bias dependence as Padovani and Stratton's expression but differs in magnitude by a factor $2(qE_{00}/kT)^{\frac{1}{2}}$. If $0.1 < (qE_{00}/kT) < 1$, the difference between the two expressions is not significant.

At higher doping concentrations field emission may occur. Unlike the case of forward bias, field emission can take place in non-degenerate semiconductors because it involves tunnelling of electrons from the metal into the semiconductor and the metal is always degenerate. The analyses of Padovani and Stratton and of Honeisen and Mead both predict that the transition from thermionic-field to field emission occurs when $(\varphi_b^{\frac{1}{2}}kT/qE_{00}V_d^{\frac{1}{2}}) \approx 1$ where $V_d = \varphi_b+V_r-\xi$. The boundaries between the two regimes are shown in Fig. 3.16, which is a plot of V_d against kT with E_{00} as a parameter. The barrier height φ_b has been taken as 1 eV. For silicon, donor concentrations in excess of about 5×10^{24} m^{-3} are necessary if field emission is to occur at room temperature. The expression given by Padovani and Stratton for the J/V characteristic under field-emission conditions is complicated, but in the very low temperature limit it simplifies to

$$J = A^*\left(\frac{qE_{00}}{k}\right)^2 \frac{(\varphi_b+V_r-\xi)}{\varphi_b} \exp\left\{-\frac{2\varphi_b^{3/2}}{3E_{00}(\varphi_b+V_r-\xi)^{\frac{1}{2}}}\right\}. \tag{3.33}$$

Fig. 3.17 shows the forward and reverse characteristics of a gallium arsenide diode with a donor concentration of 2×10^{24} m^{-3} (Padovani 1971), the reverse current being due to field emission. At low bias voltages the reverse current exceeds the forward current as was predicted originally by Wilson (1932).

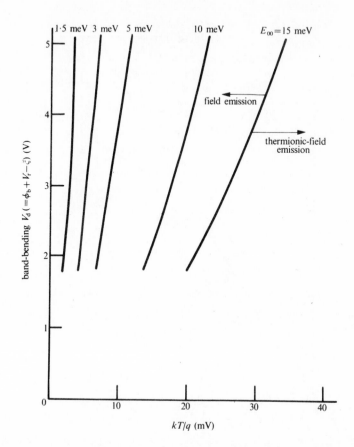

FIG. 3.16. Boundaries between field and thermionic-field emission on a
plot of band bending V_d against temperature, with E_{00} as a parameter.
The curves involve approximations which are valid only if $v_r > \varphi_b$, i.e. if
V_d is greater than about 2 V (derived from Padovani and Stratton's Fig.
12).

Tunnelling is one of the most common causes of 'soft' reverse characteristics. It is particularly important near the edge of the metal contact,
because the crowding of the field lines causes an increase in the field
strength which decreases the barrier width and also exaggerates the image-
force lowering. The effect is accentuated if the surface of the semiconductor adjacent to the metal is accumulated due to the presence of positive
surface charges as this makes the barrier at the edge even thinner (Yu and
Snow 1968).

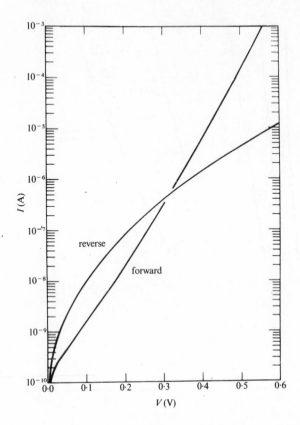

FIG. 3.17. Forward and reverse characteristics of a gallium-arsenide
Schottky diode having $N_d = 2 \times 10^{24}$ m^{-3} (Padovani 1971. Copyright Academic
Press Inc.)

Edge effects may be minimized by using a method of surface preparation
which does not cause accumulation (Smith 1968) and almost totally eliminated by using a guard-ring. The simplest type of guard-ring is a concentric
metal ring which is maintained at the same potential as the main contact
(Padovani 1968, Tove et al. 1973), but this is effective only if the
separation between the guard-ring and main contact is comparable with the
depletion width. This restricts its use to high-resistivity material. A
more effective method is to use a diffused p-type ring under the edge of
the contact (Lepselter and Sze 1968) as shown in Fig. 3.18. The presence of
the p-type material eliminates the high field at the edge and places a
reverse-biased p–n junction in parallel with the Schottky barrier. The

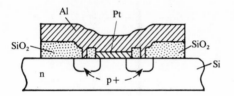

FIG. 3.18. Schottky diode with p-type guard-ring (Lepselter and Sze 1968).

p—n junction can be made to have good reverse characteristics, and the resulting diode is then very nearly ideal with a reverse characteristic determined by image-force lowering. Hole injection from the p—n junction is normally greater than from the Schottky diode alone, and if it is important to minimize hole injection two p—n junctions back-to-back may be used (Saltich and Clark 1970). Edge effects can also be minimized by using a mesa structure (Gutknecht and Strutt 1974).

If edge effects are avoided, a Schottky diode will eventually exhibit breakdown which, depending on the donor density, is due either to thermionic-field emission or to impact ionization causing an avalanche of electron—hole pairs. The dependence of the avalanche breakdown voltage on donor density is the same as for a planar p—n junction (Sze and Gibbons 1966, Okuto and Crowell 1974), and the transition from avalanche breakdown to thermionic-field emission takes place at donor densities between 5×10^{23} m^{-3} and 10^{24} m^{-3} for silicon at room temperature (Chang and Sze 1970). Coe (1971) has reported the fabrication from 25 Ω-cm silicon of aluminium—silicon diodes with multiple guard-rings which exhibit reverse breakdown voltages in excess of 1 kV.

3.6.3. *Generation in the depletion region*

If the effects of tunnelling and image-force lowering are reduced to negligible proportions by the use of a low donor density, there may be an appreciable reverse current due to the generation of electron—hole pairs in the depletion region. This is the inverse of process (c) (§ 3.1) and gives rise to a current-density component $J_g = q n_i w / 2 \tau_r$, where w is the width of the depletion region and τ_r the lifetime within the depletion region. J_g increases with reverse bias because w is proportional to $(V_{do} + V_r)^{\frac{1}{2}}$. Like the inverse process, generation current is most important in high barriers and in low-lifetime semiconductors, and is more pronounced at low temperatures than at high temperatures because it has a lower activation energy than the thermionic-emission component. It is a frequent

cause of the lack of saturation of the reverse current in gallium-arsenide
Schottky diodes.

3.7. TRANSIENT EFFECTS

It is pertinent to inquire at this stage whether there are any transient
effects in Schottky diodes which arise from a mechanism analogous to
minority-carrier storage in p—n junctions. In a p—n junction, the application of forward bias results in the injection of minority carriers (predominantly electrons into the p-type side if the donor density greatly
exceeds the acceptor density). If the bias is suddenly reversed, these injected minority carriers must be cleared away before the diode assumes a
high-resistance state, and a large current momentarily flows in the reverse
direction. The charge associated with the injected minority carriers is
equal to $J\tau_{re}$ per unit area, where J is the forward current density and τ_{re}
is the recombination time of electrons in the p-type region (see, for
example, Sze 1969).

The analogue of this process in a Schottky diode is not the hole injection discussed in § 3.5, which refers to the injection of those carriers
(holes) which contribute very little to the current, but rather the injection of electrons into the metal. As we saw in § 3.2.1 the electrons
injected into the metal are 'hot' in the sense that they possess much more
energy than corresponds to thermal equilibrium, and under the application
of reverse bias they can diffuse back into the semiconductor until they
have lost so much energy that they can no longer surmount the barrier. The
time taken for this to happen is effectively equal to the mean free time
for electron—electron collisions in the metal, since in such a collision
the hot-electron loses on average about half its excess energy (Loveluck
and Rhoderick 1967); this time plays a role equivalent to the electron-
recombination time in a p—n junction. The mean free path for electron—
electron collisions when the 'hot' electron has 1 eV excess energy above
the Fermi level is about 500 Å (Ritchie and Ashley 1965) and the mean free
time about 10^{-14} s. This is negligibly small, and minority-carrier-storage
effects arising from this mechanism have not been observed in Schottky
diodes.

It is natural to ask whether there may be an observable recovery time
in a Schottky diode associated with the charge due to the injected holes
because, although the hole current is very much smaller than the electron
current, the hole recombination time in the semiconductor τ_{rh} ($\sim 10^{-6}$ s in
a typical silicon diode) is many orders of magnitude greater than the

electron—electron mean free time in the metal. The injected-hole charge per unit area is given by $J_h \tau_{rh}$ which is equal to $\gamma_h J \tau_{rh}$, where J is the total forward current density and γ_h the hole-injection ratio given by eqn (3.30). The charge storage arising from hole injection can be described in terms of an effective recovery time τ_h equal to the injected charge per unit area divided by the total current density J, which is equal to $\gamma_h \tau_{rh}$. Although τ_{rh} may be comparatively long, γ_h is so small that the recovery time due to injected holes is usually not more than about 10^{-11} s and is negligible for most purposes. At high current densities τ_h rises because of the increase in γ_h and depends in a complicated way on the boundary conditions at the back (ohmic) contact as has been discussed by Scharfetter (1965). In practice, the recovery time of a Schottky diode is determined by circuit considerations rather than by any intrinsic electronic process associated with the conduction mechanism.

3.8. THE EFFECT OF AN INTERFACIAL LAYER

Unless they are manufactured by cleaving the semiconductor in an ultra-high vacuum, Schottky diodes nearly always have a thin oxide layer between metal and semiconductor. This interfacial layer may be considered to be an insulator, even though it may be so thin (of the order of 10 Å) that it does not possess the band-structure characteristic of a thick oxide. An idealized band diagram for such an imperfect Schottky barrier is shown in Fig. 3.19.

The insulating layer has three effects:

(i) Because of the potential drop in the layer, the zero-bias barrier height is lower than it would be in an ideal diode[†].

(ii) The electrons have to tunnel through the barrier presented by the insulator so that the current for a given bias is reduced in a manner equivalent to a reduction in A^{**}.

(iii) When a bias is applied, part of the bias voltage is dropped across the insulating layer so that the barrier height φ_b is a function of the bias voltage (see § 2.3.4). The effect of this bias dependence of the barrier height is to change the shape of the current/voltage characteristic in a manner which can be described in terms of an ideality factor n defined by eqn (3.14).

[†] By 'the barrier height' we mean φ_b; i.e. the maximum height of the bottom of the conduction band in the semiconductor relative to the metal Fermi level. This is distinct from the barrier presented by the insulating layer.

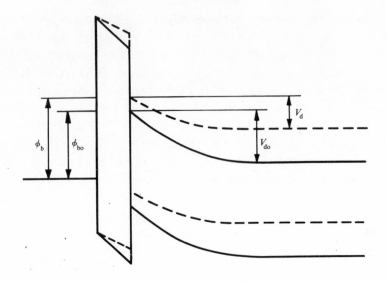

FIG. 3.19. Schottky barrier with interfacial layer. ——— zero bias, -----
forward bias.

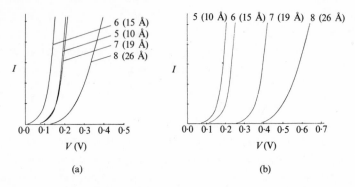

FIG. 3.20. (a) Forward I/V characteristics of Schottky diodes with oxide
interfacial layers of various thicknesses. (b) Same data as (a) normalized
to a common diffusion potential (Card and Rhoderick 1971a).

Fig. 3.20(a) shows the forward I/V characteristics of some gold—silicon
Schottky diodes prepared by Card (Card and Rhoderick 1971a) in which a
thin film of SiO_2 was intentionally grown on the silicon before the metal
was deposited. It is apparent that the current decreases with increasing
film thickness due to the effects mentioned above. Effect (i) can be eliminated by measuring V_{d0} independently and normalizing the curves to a common zero-bias diffusion potential. Fig. 3.20(b) shows the curves so

normalized; these now exhibit the combined effects of (ii) and (iii). The
reduction of current due to tunnelling through the oxide layer is clearly
visible, but there is also a departure from ideal behaviour which may be
described by an n-value which increases with increasing oxide thickness.
The dependence of the n-value on the oxide thickness may be explained in
terms of the analysis of § 3.2.4. According to eqn (3.14) we have

$$\frac{1}{n} = 1 - \frac{\partial \varphi_e}{\partial V}.$$

We shall ignore image-force lowering so that the bias dependence of φ_e is
contained entirely in φ_b, and hence $\partial \varphi_e / \partial V = \partial \varphi_b / \partial V$. From eqn (2.17),
after some manipulation and making use of eqn (2.9), we find that

$$\frac{\partial \varphi_b}{\partial V} = \frac{\alpha q N_d}{\varepsilon_s \mathcal{E}_{max} + \alpha q N_d}$$

so that

$$\frac{1}{n} = \frac{\varepsilon_s \mathcal{E}_{max}}{\varepsilon_s \mathcal{E}_{max} + \alpha q N_d}$$

or

$$n = 1 + \frac{\alpha q N_d}{\varepsilon_s \mathcal{E}_{max}}$$

$$= 1 + \frac{\alpha}{w}$$

$$= 1 + \frac{\delta \varepsilon_s}{w(\delta q D_s + \varepsilon_i)} \tag{3.34}$$

where w is the width of the depletion region. We have made use of the
result (from the depletion approximation) that $\mathcal{E}_{max} = q N_d w / \varepsilon_s$. A similar
result, but omitting the effect of surface states, has been obtained by
Strikha (1964). Card and Rhoderick (1971a) have extended this analysis to
the case where the interface states are in equilibrium with the bulk of the
semiconductor rather than with the metal. They suppose that the interface
states can be divided into two groups, such that all the states in one
group (density D_{sa}) are in equilibrium with the metal while those in the
other group (density D_{sb}) are in equilibrium with the semiconductor. The
general expression for n becomes

$$n = 1 + \frac{(\delta/\varepsilon_i)\{(\varepsilon_s/w)+qD_{sb}\}}{1 + (\delta qD_{sa}/\varepsilon_i)}.$$ (3.35)

The dependence on D_{sa} and D_{sb} can be understood physically because states in equilibrium with the metal tend to hold φ_b constant (i.e. to reduce n) while states in equilibrium with the semiconductor tend to hold V_d constant (i.e. to increase n). Eqn (3.35) has been used by Walker (1974) as the basis of a method of measuring surface state densities from I/V characteristics. Generally speaking, if $\delta < 30$ Å most of the states are in equilibrium with the metal, while if $\delta > 30$ Å most of the states are in equilibrium with the semiconductor. Films of thickness of the order of 20 Å usually lead to values of n in the range 1.3–1.5. The transition coefficients of the films used in Card's experiments were always greater than was expected on the basis of the known band structure of SiO_2, being typically of the order of 10^{-3} for a 20 Å film. Gray (1973) has suggested that this may be due to the presence of pinholes.

Another effect of an interfacial layer is to increase the minority-carrier-injection ratio under forward bias. For very thin layers, the effect arises predominantly because the electron current is limited by thermionic emission and is therefore proportional to the probability of electrons tunnelling through the oxide layer, whereas the hole current is controlled by diffusion in the neutral region of the semiconductor and is relatively unaffected by the presence of the interfacial layer. For thicker layers there is a major realignment of the bands in the semiconductor with respect to the Fermi level in the metal, with the result that far more holes are able to tunnel from the metal into the semiconductor. The effect has been investigated by Card and Rhoderick (1973) in the context of gold–silicon diodes, and exploited by Livingstone, Turvey, and Allen (1973) and Haeri and Rhoderick (1974) to improve the injection efficiency of electroluminescent diodes. Values of γ of about 0.1 have been obtained in this way.

In the reverse direction, the presence of an interfacial layer causes the effective barrier height to decrease with increasing bias, so that the reverse current does not saturate. The effect is equivalent formally to a reduction in barrier height of the form $\Delta\varphi_b = \alpha\mathcal{E}_{max}$ (see § 2.3.2 and § 3.6.1). Card and Rhoderick (1971b) have shown that, because of the reduction in barrier height, the reverse current of a diode with a fairly thick interfacial layer may actually be greater than that of a diode with a very thin layer (see Fig. 3.21). Although the electrons have to tunnel through

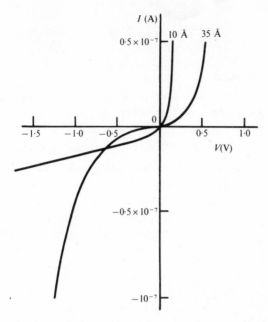

FIG. 3.21. Reverse characteristics of Schottky diodes with oxide inter-facial layers of various thicknesses (Card and Rhoderick 1971b).

the barrier presented by the insulator, this barrier is very thin and is over-compensated by the reduction in the Schottky barrier. This effect is a common cause of the 'soft' reverse characteristics of diodes prepared under poor vacuum conditions or with a contaminated semiconductor surface. Thanailakis and Northrop (1971) have found that in aluminium–germanium diodes with an observable interfacial layer the lowering of the barrier under reverse bias can exhibit a long time constant of the order of tens of minutes or even hours; this can be explained in terms of a model involv-ing slow interface states.

3.9. THE 'T_0' EFFECT

If the departure of n from unity arises from image-force lowering or from interface effects, n should be independent of temperature, but if it is due to thermionic-field emission or to the effect of recombination in the depletion region, n will be temperature dependent. The majority of Schottky diodes exhibit n values which depend on temperature. Padovani and Sumner (1965) and Padovani (1967) have shown that for some Schottky diodes made from silicon or gallium arsenide the J/V characteristic can be well represented by the equation

$$J = AT^2 \exp\{-\varphi_{b0}/k(T+T_0)\}[\exp\{qV/k(T+T_0)\}-1]$$

where T_0 is a parameter which is independent of temperature and voltage over a wide range of temperatures. This is equivalent to writing $n = 1 + (T_0/T)$, so the experimental results imply a very special sort of temperature dependence of n. Padovani (1971) states that, for a total of 25 diodes made on the same slice of gallium arsenide, the values of T_0 varied from 10 K up to 100 K. Moreover, other diodes have been reported (e.g. Arizumi and Hirose 1969; Gol'dberg, Posse, and Tsarenkov 1975) which exhibit almost ideal J/V characteristics with $T_0 \approx 0$. The effect is evidently not an intrinsic property of ideal Schottky barriers but an artefact introduced by the fabrication process. Various attempts have been made to explain such a temperature dependence in terms of tunnelling (Crowell and Rideout 1969), particular distributions of interface states (Saxena 1969, Levine 1971, Rhoderick 1975), and a non-uniformly doped surface layer (Padovani 1971, Crowell 1977). It is quite possible that more than one of these mechanisms may operate simultaneously.

It is curious that for these diodes the temperature dependence of the term involving φ_{b0} should apparently have the same form as that of the term involving V. Although all the mechanisms which have been invoked to explain the form of the current/voltage characteristic (i.e. n values greater than unity) also affect the zero-bias barrier height, they do not do so in the same way. There is also a temperature dependence of φ_{b0} which arises from the variation of the band gap with temperature, but this makes little contribution to the temperature dependence of the term involving V.

4. The capacitance of a Schottky barrier

4.1. THE CAPACITANCE OF AN IDEAL DIODE UNDER REVERSE BIAS

The depletion region of a Schottky barrier behaves in some respects like a parallel-plate capacitor. It is important to know what factors determine its capacitance, not only because reverse-biased diodes are used in practice as variable capacitors (varactors), but also because measurements of the capacitance under reverse bias can be used to give information about the barrier parameters.

4.1.1. *The general case*

Consider the ideal Schottky diode (without any interfacial layer) with a band diagram as shown in Fig. 4.1(a). The semiconductor is assumed to be n-type. The solid lines refer to a reverse bias V_r and the broken lines to a larger reverse bias $V_r + \Delta V_r$. If the reverse bias is increased from V_r to $V_r + \Delta V_r$, electrons in the conduction band of the semiconductor recede further from the metal and increase the width of the depletion region from w to $w + \Delta w$. At the same time, if there is a significant concentration of holes in the semiconductor immediately adjacent to the metal (as will happen if the surface of the semiconductor is inverted*), the hole concentration will decrease because the hole quasi-Fermi level coincides with the Fermi level in the metal (Rhoderick 1972). The change in the charge in the depletion region gives rise to capacitance. In addition, there is a parallel conductance due to the reverse current of the diode, so the equivalent circuit consists of a capacitor in parallel with a resistor.

There are three sources of charge in the barrier region: the positive charge Q_d due to the uncompensated donors in the depletion region, the positive charge Q_h due to the extra holes in the valence band, and the negative charge Q_m due to electrons on the surface of the metal. It is not immediately apparent which of these charges have to be taken into account in calculating the capacitance, but the question can be resolved by considering the various contributions to the current through the depletion region. When the bias changes, the total current through the depletion

---

*The surface is said to be inverted if the barrier height is so large that the region of the semiconductor adjacent to the metal is p-type.

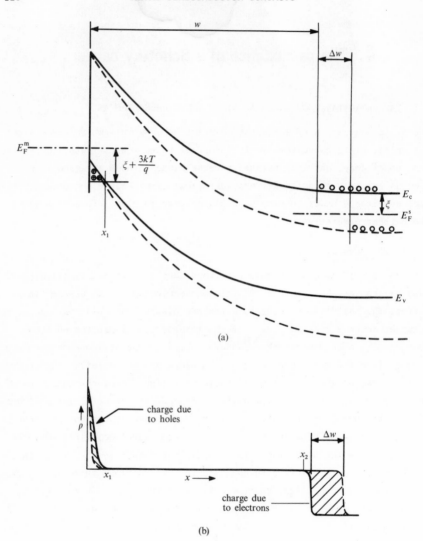

FIG. 4.1. (a) Schottky barrier under reverse bias. ——— bias voltage V_r, ----- bias voltage $V_r + \Delta V_r$. (b) Charge density due to mobile carriers (charge on surface of metal not shown).

region consists of a conduction current plus a displacement current. The displacement current density J_d (= $\varepsilon_s \partial \mathscr{E}/\partial t$) arises because the electric field in the depletion region increases with time as the bias increases. The conduction current consists of two parts – a current density J_{c1} due to the drift and diffusion of electrons injected over the barrier from the metal, and a current density J_{c2} which arises because there is a flow of electrons and holes out of the depletion region as the negative bias increases. J_{c1} exists even if the bias is constant in time and constitutes the reverse current of the diode; it is independent of position throughout the depletion region. The component J_{c2} exists only if the bias is changing. If the bias increment ΔV_r varies sinusoidally with time, the component J_{c1} is in phase with ΔV_r and gives rise to the parallel conductance, while the components J_{c2} and J_d are in phase quadrature with ΔV_r and give rise to the capacitance. Both J_{c2} and J_d depend on x, but in such a way that the sum $J_{c2} + J_d$ is constant.

The charge density due to the mobile carriers (electrons and holes) is shown in Fig. 4.1(b). To the right of the plane $x = x_1$, where $E_F^m - E_v(x_1) > \xi + (3kT/q)$, the hole density will be less than a few percent of N_d (assuming $N_c \approx N_v$); and to the left of the plane $x = x_2$, where $E_c(x_2) - E_c(\infty) > 3kT/q$, the electron density will be less than a few percent of N_d. Therefore within the range x_1 to x_2 the charge density is essentially equal to the charge on the donors qN_d and is independent of time. As the bias varies electrons enter or leave the depletion region from the bulk of the semiconductor and holes enter or leave from the metal, so that in the region x_1 to x_2 the component J_{c2} of the conduction current density must be zero, and the only quadrature component will be J_d. Hence the capacitive current will be equal to the displacement current $J_d = \varepsilon_s \partial \mathscr{E}/\partial t$. From the usual circuit definition of capacitance, the displacement current density can also be written as $J_d = C(\partial \Delta V_r/\partial t)*$, where C is the differential capacitance per unit area, so that

$$C\frac{\partial \Delta V_r}{\partial t} = \varepsilon_s \frac{\partial \mathscr{E}}{\partial t} .$$

Now $\mathscr{E}$ is an explicit function of V_r so we may write

$$\frac{\partial \mathscr{E}}{\partial t} = \frac{\partial \mathscr{E}}{\partial V_r} \times \frac{\partial \Delta V_r}{\partial t}$$

and therefore

---

*V_r is assumed independent of time, and the time dependence is contained in ΔV_r.

$$C = \epsilon_s \frac{\partial \mathscr{E}}{\partial V_r} \,. \tag{4.1}$$

The electric field strength $\mathscr{E}$ must be evaluated within the region x_1 to x_2, where the charge density is essentially equal to the donor charge density and is independent of bias.

The field strength $\mathscr{E}$ can easily be found by applying Gauss's theorem to a closed surface bounded by two planes parallel to the interface, one being situated between x_1 and x_2 and the other far into the interior of the semiconductor where the bands are flat and the charge on the electrons compensates the charge on the donors. If the bias varies, the charge enclosed by this surface varies because of the change in the width of the depletion region, and the variation in this charge is equal to the change in the total charge per unit area Q_d due to the uncompensated donors. Hence

$$\epsilon_s \Delta \mathscr{E} = \Delta Q_d$$

or

$$\epsilon_s \frac{\partial \mathscr{E}}{\partial V_r} = \frac{\partial Q_d}{\partial V_r}$$

and

$$C = \frac{\partial Q_d}{\partial V_r} \,. \tag{4.2}$$

This result shows that the capacitance can be calculated by considering only the charge due to the uncompensated donors and ignoring the charge due to the holes. Alternatively, since the entire barrier region must be electrically neutral because there is no electric field in the metal or the interior of the semiconductor, $Q_m + Q_h + Q_d = 0$ and we may write

$$C = - \frac{\partial (Q_m + Q_h)}{\partial V_r} \,.$$

This shows that the holes are to be regarded as 'belonging' to the metal because in Fig. 4.1(a) they enter from the left. The capacitance defined in this way is a differential capacitance per unit area; i.e. it represents the response of the depletion region to a change ΔV_r in the reverse bias. It can be measured experimentally by superposing a small a.c. voltage on the steady d.c. bias V_r and measuring the capacitance in the normal way by means of a bridge which responds only to the a.c. component of the current.

4.1.2. *The case in which the minority carriers are negligible*

The differential capacitance can easily be expressed in terms of the diffusion voltage and donor density if the effect of the holes can be neglected. This will be the case if the barrier height is such that the top of the valence band at the interface is below the metal Fermi level by at least $\xi + (3kT/q)$; this is equivalent to taking $x_1 = 0$. In this case the electric field at the interface will be due only to the uncompensated donors and can be found from eqn (B.1) of Appendix B with p_s put equal to zero so that

$$\mathcal{E}^2_{max} = \frac{2q}{\varepsilon_s}\left\{N_d\left(V_d - \frac{kT}{q}\right) + \frac{kTN_d}{q}\exp\left(-qV_d/kT\right)\right\}$$

where V_d is the diffusion voltage associated with the reverse bias V_r. If $qV_d > 3kT$ the last term in the bracket is negligible and

$$\mathcal{E}^2_{max} = \frac{2qN_d}{\varepsilon_s}\left(V_d - \frac{kT}{q}\right).$$

From Gauss's theorem the charge due to the uncompensated donors is given by

$$Q_d = \varepsilon_s\mathcal{E}_{max} = (2q\varepsilon_sN_d)^{\frac{1}{2}}\left(V_d - \frac{kT}{q}\right)^{\frac{1}{2}}$$

so that

$$C = \frac{\partial Q_d}{\partial V_r} = \frac{\partial Q_d}{\partial V_d}$$

$$= \left(\frac{q\varepsilon_sN_d}{2}\right)^{\frac{1}{2}}\left(V_d - \frac{kT}{q}\right)^{-\frac{1}{2}}. \tag{4.3a}$$

Since $V_d = V_{d0} + V_r$, where V_{d0} is the diffusion voltage at zero bias, eqn (4.3a) may be written

$$C = \left(\frac{q\varepsilon_sN_d}{2}\right)^{\frac{1}{2}}\left(V_{d0} + V_r - \frac{kT}{q}\right)^{-\frac{1}{2}}. \tag{4.3b}$$

Eqn (4.3b) shows that a graph of C^{-2} as a function of V_r should be a straight line with a slope of $2/q\varepsilon_sN_d$ and a negative intercept $-V_I$ on the V_r axis equal to $-V_{d0} + (kT/q)$ as shown in Fig. 4.2, so that

$V_{d0} = V_I + (kT/q)$. The distance ξ of the Fermi energy below the conduction band, and hence the barrier height $\varphi_b = V_{d0} + \xi$, can be obtained by finding N_d from the slope of the line since $\xi = (kT/q)\ln(N_c/N_d)$. Eqn (B.1) is obtained with the use of Boltzmann statistics and is therefore valid only for non-degenerate semiconductors. Goodman and Perkins (1964) have shown that for degenerate semiconductors the term $-kT/q$ in eqn (4.3a) and (4.3b) must be replaced by $-2|\xi|/5$, where ξ is now negative.

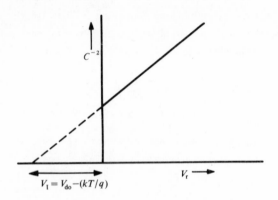

FIG. 4.2. Variation of C^{-2} with V_r for Schottky barrier.

If the term kT/q is omitted from eqn (4.3a), the result is equivalent to using the depletion approximation. In this case the relationship between C and V_d can be found very simply from first principles, because according to the depletion approximation the diffusion voltage is given by

$$V_d = \frac{qN_dw^2}{2\varepsilon_s} = \frac{Q_d^2}{2q\varepsilon_sN_d}$$

and the capacitance per unit area by

$$C = (q\varepsilon_sN_d/2V_d)^{\frac{1}{2}} = \varepsilon_s/w \tag{4.4}$$

where w is the width of the depletion region. Eqn (4.4) shows that the capacitance is the same as that of a parallel-plate capacitor with a dielectric of permittivity ε_s and thickness w. This result is easy to remember but is somewhat fortuitous. The depletion region differs from a parallel-plate capacitor in that w is a function of the bias voltage and that the charge is a space charge rather than a surface charge on the electrodes.

4.1.3. *The effect of minority carriers*

If the barrier height exceeds $E_g - \xi$, the hole density adjacent to the metal will exceed the donor density (assuming $N_c \approx N_v$), and the electric field at the interface given by eqn (B.1) will be due partly to the holes and partly to the uncompensated donors. The charge due to the holes must be subtracted from the total positive charge to find Q_d. The integral involved cannot be solved analytically, but it can be evaluated numerically using the G functions of semiconductor surface theory (see, for example, Frankl 1967). Schwarz and Walsh (1953) have made detailed calculations for germanium and have shown that, if the capacitance is expressed in the form $C \propto (V_I + V_r)^{-\frac{1}{2}}$, V_I is not equal to $V_{d0} - (kT/q)$ when the holes are taken into account but depends on the bias voltage V_r. If the barrier height exceeds $E_g - \xi$, the effect of the holes is appreciable and results in V_I being less than $V_{d0} - (kT/q)$. A plot of C^{-2} against V_r is therefore not linear, and if the low bias part of the curve is extrapolated to cut the V_r axis, the intercept will underestimate V_{d0}. The problem has also been discussed by Green (1976).

For the general case, a qualitative treatment is possible which gives a simple picture of the effect of the holes on the barrier shape and on the capacitance. It is obvious that the holes will not have much effect on the shape of the barrier unless their concentration is comparable to N_d. Remembering that the hole quasi-Fermi level ζ_h is coincident with the metal Fermi level, it is clear that the condition that $p \geqslant N_d$ is that the valence band must lie below ζ_h by less than an amount ξ, or that the bottom of the conduction band must lie above ζ_h by more than $E_g - \xi$. Suppose that this condition is satisfied to the left of the plane $x = x_3$ (Fig. 4.3). Then for $x > x_3$ the space charge is due almost entirely to the uncompensated donors and the barrier shape is parabolic. For $x < x_3$ the space charge becomes predominantly that of the holes, and because of the exponential energy dependence of p the barrier rises very steeply. The contribution of the donors to the barrier potential is as shown by the broken line in Fig. 4.3 which is simply an extrapolation of the parabolic curve. Because the barrier rises very steeply for $x < x_3$, the distance x_3 is very small and the parabolic extrapolation corresponds to a barrier height at $x = 0$ which is not very different from the height $E_g - \xi$ of the barrier at $x = x_3$. Hence the total charge due to the uncompensated donors is as if the barrier height were reduced to $E_g - \xi$, and the barrier height inferred from plots of c^{-2} against V will be less than the true value.

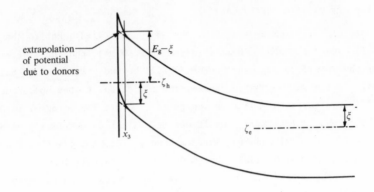

FIG. 4.3. Shape of Schottky barrier showing effect of minority carriers (holes).

A more extreme case, in which the effect of minority carriers on the capacitance is very large, has been analysed by Walpole and Nill (1971). These authors discussed what happens if the barrier height exceeds the semiconductor band gap, as is apparently the case for lead barriers on p-type lead telluride and gold barriers on p-type indium arsenide. The band diagram for such a situation is shown in Fig. 4.4 where the semiconductor is assumed to be degenerate. As in the case analysed by Schwarz and Walsh, the graph of c^{-2} against V is found not to be linear for small bias voltages although it becomes linear for large values of V. If the barrier height is regarded as a variable, the intercept V_I on the voltage axis can be shown theoretically to depend on the diffusion voltage V_{d0} as shown in Fig. 4.5. It is apparent that there is a maximum value of V_I slightly larger than the band gap E_g which is reached when V_{d0} is approximately equal to E_g. An explanation of this can be given in terms of the argument of the previous paragraph. For a degenerate semiconductor ξ is negative, so the bottom of the conduction band must lie *below* the metal Fermi level if the density of minority carriers (electrons in the p-type case) is to exceed the acceptor density. The value of the capacitance is therefore as if the barrier height were approximately equal to $E_g + |\xi|$, and an increase in true barrier height beyond this value does not result in any increase in V_I. If the voltage intercept is found experimentally to be equal to this maximum value, one can only set a lower limit to the barrier height. It is as if the inversion region in Fig. 4.4 were part of the metal.

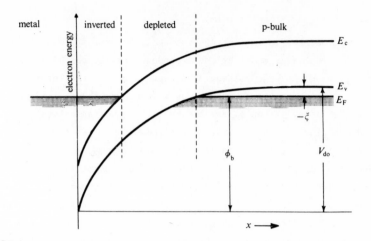

FIG. 4.4. Shape of Schottky barrier on p-type material ($\varphi_b > E_g$) (Walpole and Nill 1971).

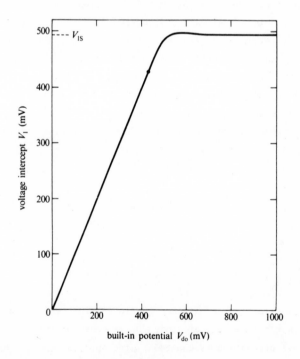

FIG. 4.5. Dependence of V_I on V_{d0} for p-type InAs at 4.2 K. E_g = 427 mV, $p = 6.3 \times 10^{16}$ cm^{-3} (Walpole and Nill 1971).

4.2. THE EFFECT OF AN INTERFACIAL LAYER

If a Schottky diode has an insulating interfacial layer, its capacitance is altered because the layer modifies the dependence of the charge in the depletion region on the bias voltage. The capacitance of the interfacial layer is effectively in series with the capacitance of the depletion region, but because the latter is non-linear the overall capacitance is a rather complicated function of the parameters involved.

The effect of an interfacial layer on the capacitance has been analysed in some detail by Cowley (1966) and by Crowell and Roberts (1969), in terms of the Bardeen model discussed in § 2.3. The analysis is easiest if the layer is sufficiently thin for the occupation of any interface states which may exist at the insulator–semiconductor interface to be determined by the Fermi level in the metal; this will be the case if the layer is not more than about 30 Å thick. The interface states are then emptied or filled by the tunnelling of electrons from the metal and make no direct contribution to the capacitance for the same reason that the minority carriers make no direct contribution in the situation discussed in § 4.1.3. The capacitance therefore is still given by $\partial Q_d/\partial V_r$. The interface states may, however, affect the capacitance indirectly because they may modify the way in which the charge due to the uncompensated donors depends on the bias voltage. This has been analysed by Cowley who showed that the graph of C^{-2} against V_r remains linear with a slope of $2/q\varepsilon_s N_d$, as in the ideal case discussed in § 4.1.2 where there is no interfacial layer. However, the intercept V_I' on the V_r axis is modified and is given by

$$V_I' = V_{d0} - \frac{kT}{q} + \alpha\left(\frac{2qN_d}{\varepsilon_s}\right)^{\frac{1}{2}}\left(V_{d0} - \frac{kT}{q}\right)^{\frac{1}{2}} + \frac{qN_d\alpha^2}{2\varepsilon_s} \qquad (4.5)$$

where

$$\alpha = \frac{\delta\varepsilon_s}{\varepsilon_i + q\delta D_s}.$$

In deriving this result it was assumed that α remains constant as V_r changes; i.e. that the density of states D_s is constant for those interface states whose occupation is changed by the application of bias, and also that the effect of the holes can be neglected. We can write eqn (4.5) in the form

$$V'_I = \cdot \varphi_{b0} - \xi - \frac{kT}{q} + \varphi_1^{\frac{1}{2}}\left(V_{d0} - \frac{kT}{q}\right)^{\frac{1}{2}} + \frac{\varphi_1}{4},$$

where φ_{b0} is the zero-bias barrier height and $\varphi_1 = 2\alpha^2 q N_d / \varepsilon_s$. If we were to deduce the barrier height from the C/V data in the usual way by adding $\xi + kT/q$ to V'_I, we should obtain the value

$$\varphi_{b0} + \varphi_1^{\frac{1}{2}}\left(V_{d0} - \frac{kT}{q}\right)^{\frac{1}{2}} + \frac{\varphi_1}{4} = \varphi_b^0 + \frac{\varphi_1}{4}$$

from eqn (2.16a). Since in all practical cases φ_1 is less than the flat-band barrier height φ_b^0 by about two orders of magnitude, the C/V method gives essentially φ_b^0. If we take the case considered at the end of § 2.3.4 with $\delta = 20$ Å and $N_d = 10^{22}$ m^{-3}, the value of α is about 60 Å if D_s does not exceed about 10^{17} ev^{-1} m^{-2}. In this case the difference between φ_{b0} and φ_b^0 would be about 0.02 eV, which is about equal to the experimental error of a careful measurement. If δ is increased to 30 Å and N_d to 10^{23} m^{-3}, the difference would be about 0.1 eV which is quite significant. The error would be reduced by a high density of surface states since these reduce α and tend to lock the height of the barrier relative to the Fermi level in the metal.

Cowley's analysis which leads to eqn (4.5) assumes that the population of the interface states can vary sufficiently rapidly for them to be in instantaneous equilibrium with the metal. This is equivalent to assuming that their time constants are much shorter than the inverse of the measuring frequency. Such an assumption is consistent with the assumption that the interface states are in equilibrium with the metal since, if the interfacial layer is thin enough to enable good communication between the interface states and the metal, their time constants will be very short and they should easily be able to follow a measuring frequency of around 100 kHz.

Cowley has compared his theoretical analysis with some experimental results on gallium-phosphide Schottky diodes made by evaporating the metal either in a conventional vacuum system fitted with an oil-diffusion pump or in a system fitted with an ion pump capable of obtaining a vacuum of 10^{-8} Torr. In the latter he found good agreement between the values of the barrier heights deduced from the C/V data and those obtained by the photoelectric method, whereas in the former the value deduced from the C/V characteristics was consistently higher than that obtained photoelectrically by as much as 0.5 eV. Cowley explained the discrepancy by supposing that in the oil-pumped system oil vapour condensed on the semiconductor before the metal was deposited, so that an interfacial layer was formed, and that the better vacuum conditions obtainable in the ion-pumped system resulted in negligible contamination. But although the results could be described qualitatively in terms of the above model, the magnitude of the discrepancy was much greater than could be explained by assuming the interface states to be in equilibrium with the metal. Cowley was forced to assume that in some of his diodes the interfacial layer was so thick that the interface states were at least partially in equilibrium with the semiconductor. This is a much more complicated situation because the interface states are now filled from the semiconductor, and the capacitance is given by $\partial(Q_d + Q_{ss})/\partial V_r$. The case is similar to that of an MOS capacitor, and the graph of C^{-2} against V_r is generally not linear.

4.3. NON-UNIFORM DONOR DISTRIBUTION

We saw in § 4.1 that, for an ideal Schottky diode with a uniform concentration of donors, the plot of C^{-2} against V is linear with a slope equal to $2/\varepsilon_s q N_d$, and in § 4.2 it was seen that this remains true in the presence of a thin interfacial layer. This result is frequently made use of as a means of measuring the donor density. If the semiconductor is compensated, then N_d must be replaced by $N_d - N_a$. We shall now show that, if the donor density is non-uniform, the plot of C^{-2} against V is not linear but its slope at any particular bias voltage is inversely proportional to the donor density at the edge of the depletion region.

Suppose that the donor distribution is non-uniform and that N_d varies with the distance x into the semiconductor as shown in Fig. 4.6(a). Let us assume that we can use the depletion approximation, according to which the electron density is negligible compared with $N_d(x)$ throughout the depletion region and rises abruptly to a value equal to $N_d(w)$ at the edge of the depletion region of thickness w at a reverse bias V_r. The charge density will then be equal to $q N_d(x)$ between $x = 0$ and $x = w$, and zero beyond $x = w$. The electric field strength $\mathscr{E}$ will be zero beyond $x = w$, and from Gauss's theorem the field strength at the position x' will be equal to the total charge between the planes $x = x'$ and $x = w$ multiplied by ε_s^{-1}; i.e. it will be proportional to the area under the $N_d(x)$ curve between x' and w. The variation of $\mathscr{E}$ with x will be as shown by the solid line in Fig. 4.6(b), and the total drop in potential across the depletion region will be equal to the area under this curve. Suppose the reverse bias is increased to

$V_r + \Delta V_r$ resulting in an increase in the width of the depletion region from w to $w + \Delta w$. An additional amount of charge $qN_d(w)\Delta w$ will be 'uncovered' at the edge of the depletion region, and the field strength at any point within the depletion region will be increased by an amount $qN_d(w)\Delta w/\varepsilon_s$ as shown by the broken line in Fig. 4.6(b). The drop in potential across the depletion region, which is now equal to the area under the broken line, has clearly been increased by an amount $w(qN_d(w)\Delta w/\varepsilon_s)$ and this must equal the increase in bias voltage ΔV_r, so that $\Delta w/\Delta V_r = \varepsilon_s/wqN_d(w)$. The increase in the charge due to the uncompensated donors (ΔQ_d) is $qN_d(w)\Delta w$, and the capacitance can be found by allowing ΔV_r to become infinitesimally small so that

$$C = \frac{\Delta Q_d}{\Delta V_r} = \frac{\varepsilon_s}{w} .$$

In other words, as with a uniform distribution of donors, the capacitance is equal to that of a parallel-plate capacitor of thickness w and permittivity ε_s. If C^{-2} is plotted as a function of V_r, the slope of the curve will be

$$\frac{\partial C^{-2}}{\partial V_r} = \frac{2w}{\varepsilon_s^2}\left(\frac{\partial w}{\partial V_r}\right) = \frac{2}{\varepsilon_s qN_d(w)} . \tag{4.6}$$

This is the same result as for a uniform distribution of donors, except that $N_d(w)$ is the donor density at the edge of the depletion region. Hence, if w is varied by changing the reverse bias, the slope of the graph of C^{-2} against V_r will give N_d at $x = w$. The value of w corresponding to a particular value of V_r can be obtained from the capacitance through the relationship $C = \varepsilon_s/w$.

The result of this analysis forms the basis of a technique which is extensively used for determining the doping profile of semiconductors, and a number of instruments for the automatic determination of impurity profiles from C/V measurements have been developed (see, for example, Baxandall, Colliver, and Fray 1971). The principle of the latter is to use a signal of frequency about 100 kHz to measure C, while the reverse bias is modulated at a lower frequency around 1 kHz so that $\partial C/\partial V_r$ can be obtained directly. From C and $\partial C/\partial V_r$ it is possible to determine $\partial C^{-2}/\partial V_r$ and w automatically. In practice the Schottky barrier is formed either by evaporating a metal such as gold onto a suitably etched surface or by using a mercury contact contained in a capillary tube. In semiconductors like

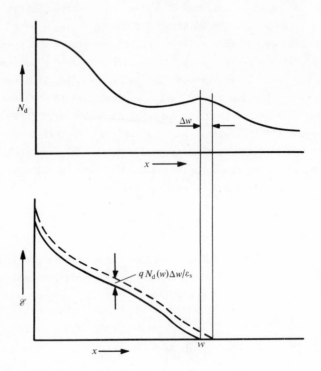

FIG. 4.6. (a) Non-uniform donor distribution, and (b) resulting electric-field distribution in depletion region.

indium phosphide which form low barriers, the high parallel conductance may cause measurement difficulties. This may be overcome by deliberately creating a thin interfacial layer to reduce the reverse current; for example, by oxidizing the surface of the semiconductor. Provided the layer is thin enough for any surface states which may be present to be in equilibrium with the metal, $\partial C^{-2}/\partial V_r$ will not be affected (see § 4.2). There are many possible sources of error which must be taken into account. Goodman (1963) has discussed general causes of error in C/V measurements, and Smith and Rhoderick (1969) have considered the effect of methods of surface preparation. Generally speaking, if the impurity profiles obtained from C/V data are to be reliable, the I/V characteristics of the Schottky diode should be not very far from ideal with n values less than about 1.1.

A more fundamental difficulty in applying the capacitance technique to the measurement of impurity profiles arises if the gradient of the impurity density is very large and the mean carrier density is low; one cannot then

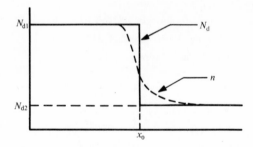

FIG. 4.7. Electron concentration n resulting from an abrupt change in donor density N_d.

be sure of the existence of charge neutrality even in the bulk of the semi-conductor. We saw in § 4.1 that the capacitance is given by $\partial Q_d/\partial V_r$, where Q_d is the charge due to the uncompensated donors. If V_r changes, Q_d changes owing to the movement of the mobile carriers (electrons if the semiconduc-tor is n-type) in and out of the depletion region and not because the donors move. The value of $\partial C^{-2}/\partial V_r$ given by eqn (4.6) therefore strictly should contain $n(w)$ in the denominator rather than $N_d(w)$, where $n(w)$ is the electron density at the edge of the depletion region. If the donor density is uniform, the bulk of the semiconductor is electrically neutral so that $n = N_d$ (assuming complete ionization of the donors) and eqn (4.6) is cor-rect. As an example of a case in which eqn (4.6) would not be correct, suppose that the donor density suddenly changes from a value N_{d1} for $x < x_0$ to a value N_{d2} for $x > x_0$, where $N_{d1} > N_{d2}$. The electron density does not fall abruptly from N_{d1} to N_{d2} at $x = x_0$ but decreases smoothly, being less than N_{d1} for some distance inside the highly doped part and greater than N_{d2} for some distance inside the weakly doped part as shown in Fig. 4.7. The region on each side of x_0 is therefore not electrically neutral, but has a net positive charge density for $x < x_0$ and a net nega-tive charge density for $x > x_0$. This non-neutral region constitutes a dipole layer which causes band bending as in a p–n junction. The distance over which the electron density differs significantly from the donor den-sity is approximately equal to the Debye length $L_D = (\varepsilon_s kT/q^2 n)^{\frac{1}{2}}$. The space-charge region therefore extends further into the more weakly doped side. If $N_{d2} \sim 10^{21} \text{ m}^{-3}$, $L_{D2} \sim 0.1 \text{ μm}$, and an attempt to produce an impurity profile for such a distribution from C/V data would yield the electron density rather than the donor density and would show the transition spread over $\sim 0.1 \text{ μm}$ instead of being abrupt. Impurity distributions in which the

density changes appreciably in distances of less than 0.1 μm cannot therefore be investigated by this technique if the minimum donor density is less
than about 10^{21} m^{-3}. A fuller discussion of this point has been given by
Kennedy, Murley, and Kleinfelder (1968).

4.4. THE EFFECT OF DEEP TRAPS

By a 'deep trap' we mean a localized electron state in the bulk of the
semiconductor which is so far removed from the conduction or valence bands
that it is not ionized at room temperature. In this respect it differs from
shallow donors and acceptors which are usually completely ionized. If the
energy level of such a trap varies with respect to the Fermi level as, for
example, in the depletion region of a Schottky barrier where the bands are
bent, the occupation of the level (and therefore its state of charge) will
also vary. Because the degree of band bending depends on the application of
a bias voltage, the state of charge of the trap will depend on the bias and
as a result the capacitance will be affected. The effect of traps on the
capacitance of a Schottky barrier affords a very convenient method of
detecting and characterizing an extremely low concentration of defects.

4.4.1. *The population of deep traps under reverse bias*

Consider a Schottky barrier in an n-type semiconductor which contains a
single type of deep trap of energy E_t as shown in Fig. 4.8. The traps are
assumed to be distributed uniformly throughout the semiconductor. In the
absence of bias, the traps will be occupied by electrons if they lie below
E_F (i.e. to the right of $x = a$ in Fig. 4.8) and will be empty above E_F,
assuming the zero-temperature approximation to the Fermi—Dirac function. If
a reverse bias is applied (Fig. 4.9(a)), the Fermi level is split into
quasi-Fermi levels for electrons and holes. As was seen in Chapter 3, it
is a very good approximation to assume that under forward bias the electron
quasi-Fermi level ζ_e is horizontal through the depletion region and coincides with the Fermi level in the semiconductor, while the hole quasi-Fermi
level ζ_h is also horizontal and coincides with the Fermi level in the metal.
This is also a fairly good approximation under reverse bias except that, as
shown by Crowell and Beguwala (1971), the electron quasi-Fermi level rises
appreciably as it approaches the metal. We shall neglect this complication
and assume both quasi-Fermi levels to remain flat. The question now to be
answered is what determines the occupation of the traps.

In the general case, the occupation of the traps is determined by
Shockley—Read—Hall statistics (see, for example, Blakemore 1962) according

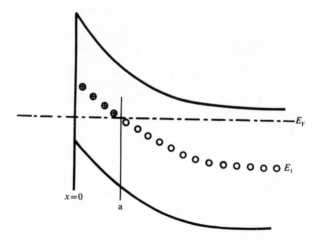

FIG. 4.8. Charge state of deep traps (zero bias). The traps are assumed to be donor-like, i.e. positively charged when empty and neutral when occupied.

to which the probability of a trap being occupied by an electron is given by

$$f = \frac{\sigma_e n \bar{v} + e_h}{\sigma_e n \bar{v} + e_e + \sigma_h p \bar{v} + e_h}. \qquad (4.7)$$

Here σ_e and σ_h are the capture cross-sections of the trap for electrons and holes, respectively, and $\bar{v}$ is the mean thermal velocity (assumed the same for electrons and holes). The probabilities of the trap emitting an electron to the conduction band or a hole to the valence band are e_e and e_h, respectively, and n and p are the concentrations of electrons and holes. Detailed balancing arguments show that e_e and e_h are related to σ_e and σ_h by

$$e_e = \sigma_e \bar{v} n_i \exp\{q(E_t - E_i)/kT\} \qquad (4.8a)$$

and

$$e_h = \sigma_h \bar{v} n_i \exp\{q(E_i - E_t)/kT\} \qquad (4.8b)$$

where n_i is the intrinsic carrier concentration and E_i the intrinsic Fermi level.

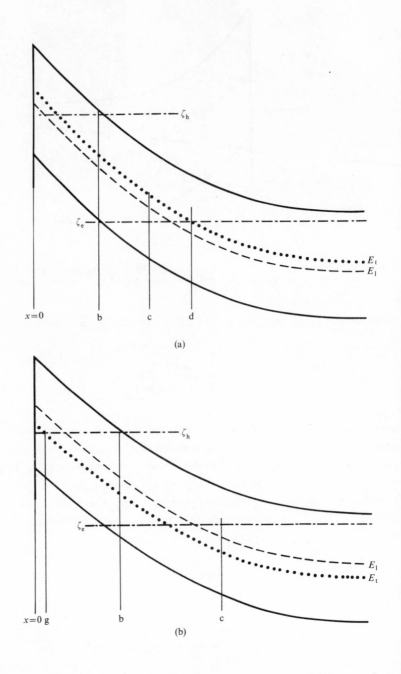

FIG. 4.9. Schottky barrier with deep traps under reverse bias (a) $E_t > E_1$ and (b) $E_t < E_1$.

It is convenient to introduce an energy E_1 defined by

$$E_1 = E_i + \frac{kT}{2q} \ln\left(\frac{\sigma_h}{\sigma_e}\right)$$

where E_i is the intrinsic Fermi level. E_1 is a parameter associated with a
particular type of trap and has the significance that, if the trap energy
lies above E_1, electron emission is more likely than hole emission while,
if the trap energy lies below E_1, the converse applies. From eqns· (4.8a)
and (4.8b) and the definition of the quasi-Fermi levels, it can easily be
shown that the following inequalities hold:

$$e_e > e_h \quad \text{if} \quad E_t > E_1 \tag{4.9a}$$

$$\sigma_e n \bar{v} > e_e \quad \text{if} \quad E_t < \zeta_e \tag{4.9b}$$

$$\sigma_e n \bar{v} > e_h \quad \text{if} \quad E_t + \zeta_e > 2E_1 \tag{4.9c}$$

$$\sigma_h p \bar{v} > e_h \quad \text{if} \quad E_t > \zeta_h \tag{4.9d}$$

$$\sigma_h p \bar{v} > e_e \quad \text{if} \quad E_t + \zeta_h < 2E_1 \tag{4.9e}$$

$$\sigma_e n > \sigma_h p \quad \text{if} \quad \zeta_e + \zeta_h > 2E_1 \tag{4.9f}$$

Because of the exponential energy dependences of the emission rates and
carrier densities, an inequality of more than $2kT/q$ in one of the energy
relationships results in a difference of an order of magnitude or more
between the corresponding emission or capture rates.

Let us first focus attention on traps in the upper half of the band gap
(Fig. 4.9a). If E_t exceeds E_1 by more than $2kT/q$, e_e will exceed e_h by at
least an order of magnitude, so that we may neglect e_h in the denominator
of eqn (4.7). We shall assume that this is the case. A further simplification may be made if $\zeta_e + \zeta_h$ exceeds $2E_1$, which occurs to the right of the
plane $x = b$ at which E_1 lies midway between ζ_e and ζ_h. For $x > b$, $\sigma_e n$ is
greater than $\sigma_h p$ from eqn (4.9f), and we may neglect $\sigma_h p \bar{v}$ in the denominator of eqn (4.7). To the right of the plane $x = c$ at which E_1 lies midway
between E_t and ζ_e, we have $E_t + \zeta_e > 2E_i$ so that $\sigma_e n \bar{v} > e_h$ from eqn (4.9c),
and we may neglect e_h in the numerator of eqn (4.7) which now reduces to

$$f \approx \frac{\sigma_e n\bar{v}}{\sigma_e n\bar{v} + e_e}.$$

It is easy to see that this represents a Fermi–Dirac distribution with a
Fermi level equal to ζ_e so that, assuming the zero-temperature approximation, the traps are fully occupied to the right of the plane $x = d$ at which
E_t coincides with ζ_e. To the left of $x = d$, f falls monotonically until for
$x < c$ we have $E_t + \zeta_e < 2E_1$, so that $\sigma_e n\bar{v} < e_h$ because of eqn (4.9c) and
f approaches the limit

$$f \approx \frac{e_h}{e_e} \approx 0$$

so that the traps are nearly all empty. Hence throughout the depletion region the population of the traps is determined by the electron quasi-Fermi
level ζ_e.

   In considering traps in the lower half of the band gap (Fig. 4.9(b)) we
shall assume that E_t lies below E_1 by more than $2kT/q$ so that $e_e < e_h$.
Again, to the right of the plane $x = b$ at which E_1 lies midway between
ζ_e and ζ_h, we have $\sigma_e n > \sigma_h p$ so that

$$f \approx \frac{\sigma_e n\bar{v} + e_h}{\sigma_e n\bar{v} + e_h} = 1.$$

To the left of $x = b$, $\sigma_e n < \sigma_h p$; in addition, since all points for which
$x < b$ also lie to the left of the plane $x = c$ at which E_1 lies midway
between E_t and ζ_e, $\sigma_e n v < e_h$ from eqn (4.9c) so that

$$f \approx \frac{e_h}{\sigma_h p\bar{v} + e_h}.$$

It is easily seen that this is a Fermi–Dirac distribution with a Fermi level
equal to ζ_h so that, in the zero-temperature approximation, the traps are
occupied to the right of the plane $x = g$ at which E_t coincides with ζ_h and
are empty for $x < g$. Hence in this case the traps are determined by the
hole quasi-Fermi level throughout the depletion region.

   One can use similar arguments to show that if $E_t \approx E_1$ so that $e_e \sim e_h$,
the traps are full to the right of the plane where E_t coincides with ζ_e

and are empty to the left of the plane where E_t coincides with ζ_h. Between these two planes the probability of occupation is approximately equal to $e_h/(e_e + e_h)$.

The assumption is often made that the occupation of the traps is determined by the electron quasi-Fermi level. It is clear from the foregoing arguments that this is only true for traps for which $E_t > E_1$. Moreover, as we have already pointed out, the quasi-Fermi level must not be assumed to be flat near to the metal.

4.4.2. *The contribution of traps to the capacitance*

Before discussing the effect of deep traps on the capacitance of a Schottky barrier, we must consider carefully how the capacitance is measured in practice. Nearly all capacitance measurements involve the use of an a.c. bridge operating with a small-amplitude test signal of frequency ω_s. To measure the capacitance as a function of reverse-bias voltage, the bias is changed at a rate which is slow compared with the test frequency ω_s. In some applications the bias consists of a d.c. component modulated by a small a.c. component which varies sinusoidally at a low frequency ω_b. Because the traps do not fill or empty instantaneously when the bias changes, their contribution to the capacitance will depend on how the two frequencies ω_s and ω_b compare with the inverse of the time constant associated with the filling of the traps. The various cases have been reviewed by Kimerling (1974). We shall consider the particular situation in which the traps lie in the upper half of the band gap and are controlled by the electron quasi-Fermi level which is assumed to be flat throughout the depletion region. It can easily be seen that in this case the trap time constant is given by $\tau_t = (e_e + \sigma_e n v)^{-1}$. Since the only traps whose occupation changes with bias are those close to the quasi-Fermi level (and for these $e_e \approx \sigma_e n \bar{v}$), it follows that the mean time constant of the traps in question is equal to $(2e_e)^{-1}$.

Case (i): $2e_e > \omega_s > \omega_b$

In this case the traps are able to follow instantaneously both the test signal and the bias variation. Suppose that the traps are donor-like (i.e. neutral when occupied, positively charged when empty), so that they are positively charged to the left of the plane $x = y$ (Fig. 4.10(a)) and neutral for $x > y$. The donors are uncompensated within the depletion region of width w, so that the variation of charge density with x is as shown by the solid line in Fig. 4.10(b). If the reverse bias is increased

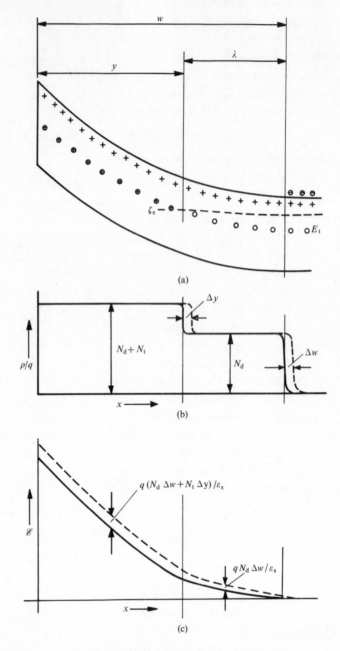

FIG. 4.10. Effect of reverse bias on deep traps. (a) Charge state, (b) charge density, and (c) field distribution. ——— reverse bias V_r, ----- reverse bias $V_r + \Delta V_r$.

from V_r to $V_r + \Delta V_r$, the charge distribution will become as shown by the broken line in Fig. 4.10(b), and the positive charge will increase by an amount $qN_d\Delta w$ due to the uncompensated donors* and by an amount $qN_t\Delta y$ due to the change in charge state of the traps. The increment ΔV_r includes both the test signal of frequency ω_s and the bias variation of frequency ω_b. The electric-field intensity will be as shown in Fig. 4.10(c), and the increase in the potential drop across the depletion region (ΔV_r) must be equal to $q(N_d w\Delta w + N_t y\Delta y)/\varepsilon_s$. The differential capacitance is therefore

$$C = \frac{\Delta Q}{\Delta V_r} = \frac{\varepsilon_s (N_d\Delta w + N_t\Delta y)}{N_d w\Delta w + N_t y\Delta y} . \tag{4.10}$$

Although Fig. 4.10(b) shows uniform donor and trap densities, it can easily be shown, using an argument similar to that in § 4.3, that eqn (4.10) is still correct for the case of non-uniform donor and trap distributions provided N_d is taken to be the donor density at $x = w$ and N_t the trap density at $x = y$. If the donor density is uniform, $w - y$ is given according to the depletion approximation by

$$w - y = \left\{ \frac{2\varepsilon_s (E_F - E_t)_{\text{bulk}}}{qN_d} \right\}^{\frac{1}{2}} = \lambda \tag{4.11}$$

so that $\Delta w = \Delta y$. In this case

$$C = \frac{\varepsilon_s (N_d + N_t)}{N_d w + N_t y} \tag{4.12}$$

and the effective depletion width is an appropriately weighted average of w and y. It is easily shown that in this case

$$\frac{\partial C^{-2}}{\partial V_r} = \frac{2}{\varepsilon_s q(N_d + N_t)} . \tag{4.13}$$

It is important to realize that the term N_t in the denominator is the trap density at $x = y$ and not at $x = w$. Since the effective depletion width obtained from eqn (4.12) lies between y and w, the apparent trap distribution will be distorted if it is deduced from eqns (4.12) and (4.13). If the

---

*It is assumed throughout this chapter that there are no shallow acceptors present. If there is a concentration N_a of shallow acceptors, N_d must everywhere be replaced by $N_d - N_a$.

traps are acceptor-like (neutral when empty), the expression for the capacitance is more complicated because the denominator of eqn (4.11) now contains $N_d - N_t$, and λ is not constant unless both N_d and N_t are constant, so that in general Δw is not equal to Δy.

Case (ii): $\omega_s > 2e_e > \omega_b$

In this case the traps cannot follow the test signal, and the capacitance is determined entirely by the movement of electrons at the edge òf the depletion region so that

$$C = \frac{\varepsilon_s}{w}. \tag{4.14}$$

However, the mean occupation of the traps (averaged over a time long compared with ω_s^{-1}) is able to follow the variation of the bias voltage at the frequency ω_b and affects the relationship between w and v_r. For donor-like traps, the change in w and y due to a change in the bias voltage ΔV_r is still given by

$$\Delta V_r = \frac{q}{\varepsilon_s}(N_d w\Delta w + N_t y\Delta y).$$

(ΔV_r here refers only to the change in bias voltage at frequency ω_b. It does not include the component at frequency ω_s due to the test signal.) Consequently

$$\frac{\partial C^{-2}}{\partial V_r} = \frac{2w}{\varepsilon_s^2} \times \frac{\partial w}{\partial V_r} = \frac{2}{q\varepsilon_s\left\{N_d+N_t\left(\frac{y}{w}\right)\right\}} \tag{4.15}$$

if one assumes a uniform donor density so that $\Delta w = \Delta y$. Hence, for large bias voltages when y/w tends towards unity, the apparent donor density tends towards $N_d + N_t$. If N_d is constant, eqn (4.15) can be used to infer N_t, but it must be remembered that N_t is the trap density at $x = y$ and not at the position $x = w$ given by eqn (4.14), so the apparent trap distribution inferred from eqns (4.14) and (4.15) will be distorted. Eqn (4.15) can be rewritten as

$$\frac{\partial V_r}{\partial C^{-2}} = \frac{q\varepsilon_s}{2}\left(N_d+N_t-\frac{\lambda N_t C}{\varepsilon_s}\right) \tag{4.15a}$$

where $\lambda = w - y$ and, if N_d and N_t are constant, a plot of $\partial V_r/\partial C^{-2}$ against

C gives a straight line with an intercept at $C = 0$ that gives the value of $N_d + N_t$. This was first pointed out by Senechal and Basinski (1968a). The case of acceptor-like traps is again very complicated because λ is not constant so that Δy is no longer equal to Δw.

If the test frequency ω_s is increased so that there is a gradual change from case (i) to case (ii), there will be a smooth change from the low-frequency capacitance C_{1f} given by eqn (4.12) to the high-frequency capacitance C_{hf} given by eqn (4.14). It can be shown (Zohta 1973) that at an intermediate frequency the capacitance is given by

$$C = C_{hf} + \frac{C_{1f} - C_{hf}}{1 + \omega_s^2 \tau^{*2}} \, ,$$

where τ^* is related to, but not quite the same as, $(2e_e)^{-1}$. In the transition region where $\omega_s \tau^* \approx 1$, the dispersion in capacitance is accompanied by a parallel conductance. The time constant of the trap can be determined either by measuring this conductance as a function of ω_s or by changing the temperature at constant ω_s so as to change τ^* and thereby obtain the condition $\omega_s \tau^* \approx 1$ (Losee 1975; Vincent, Bois, and Pinard 1975).

Case (iii): $\omega_s > \omega_b > 2e_e$

In this case the traps cannot follow either the test signal or the modulation of the bias voltage. The capacitance is determined entirely by the movement of electrons at the edge of the depletion region and is given by eqn (4.14); the traps do not affect the variation of depletion width with bias voltage so that

$$\frac{\partial C^{-2}}{\partial V_r} = \frac{2}{\varepsilon_s q N_d} \tag{4.16a}$$

if the traps are donor-like, and

$$\frac{\partial C^{-2}}{\partial V_r} = \frac{2}{\varepsilon_s q (N_d - N_t)} \tag{4.16b}$$

if the traps are acceptor-like. In eqn (4.16b), unlike cases (i) and (ii), N_t is the trap concentration at $x = w$.

Case (iii) includes as a special case the method invented by Copeland (1969). In Copeland's technique a current containing an a.c. component at frequency ω_s is driven through the diode, and the a.c. voltage components across the diode at frequencies ω_s and $2\omega_s$ are measured using tuned amplifiers. The component at frequency ω_s is proportional to C^{-1} and gives the depletion width w, while the component at frequency $2\omega_s$ can be shown to be proportional to N_d^{-1} where N_d is the donor density at $x = w$. This result is true whether traps are present or not. It was pointed out by Zohta (1970) that if Copeland's method is used to find N_d the result can be combined with eqns (4.13) or (4.15a) to obtain N_t for cases (i) and (ii), respectively. If N_d and N_t are both constant, the slope of a plot of $\partial V_r/\partial C^{-2}$ against C gives λ from eqn (4.15a), and the trap energy E_t can then be obtained from eqn (4.11).

It is clear from the above considerations that if traps are present the capacitance will be a complicated function of frequency*. Conversely, if the capacitance is found experimentally to be independent of the test frequency, this can be regarded as evidence for the absence of traps. If traps are present, great care must be taken in using the C/V technique to measure impurity profiles, and the only reliable method is to ensure that the test and bias modulation frequencies satisfy the conditions of case (iii). A comprehensive discussion of the problems of impurity profiling in the presence of traps has been given by Kimerling (1974).

4.4.3. Transient measurements

There is an important class of measurements in which the reverse bias on a Schottky barrier is changed instantaneously from one value to another, and the resulting change in capacitance is observed as a transient. The measurement can be made by using a capacitance meter which gives a voltage output proportional to the capacitance under test; the capacitance can then be observed as a function of time by displaying the meter output on an oscilloscope or chart recorder. The method was first used by Williams (1966) and subsequently refined by Senechal and Basinski (1968a, 1968b).

Consider a Schottky barrier containing a uniform concentration of traps N_t under a reverse bias voltage V_1. If the traps are donor-like and lie in energy above E_1, so that their population is governed by the electron quasi-Fermi level, the band diagram will be as shown in Fig. 4.11(a). Those traps which are at more than a distance y_1 from the metal will be neutral, and those that are within y_1 of the metal will be positively charged. If the differential capacitance C_1 (per unit area) is measured with a test frequency to which the traps cannot respond, C_1 will be given by $C_1 = \varepsilon_s/w_1$. Suppose now that at time $t = 0$ the reverse bias is instantaneously increased

---

*Unless the conditions are such that the traps are always full. This will be the case if $E_t < E_1$ and also $E_t < \zeta_h$ at the interface.

to V_2. Because the traps cannot change their charge state instantaneously, they will remain positively charged up to a distance y_1 from the metal at $t = 0^+$. Since the reverse bias is increased, the depletion width must increase instantaneously to $w_2(0)$, and the capacitance $C_2(0)$ measured at $t = 0^+$ will be equal to $\varepsilon_s/w_2(0)$. As time proceeds, those traps which lie above the electron quasi-Fermi level at distances between y_1 and $y_2(t)$ from the metal will lose their electrons and become positively charged, since the electron-emission rate e_e is now much greater than the electron capture rate $\sigma_e nv$. The positive charge density is thereby increased, and the depletion width will decrease. Finally, as $t \to \infty$, all the traps above the electron quasi-Fermi level will become positively charged, the depletion width will decrease to a value $w_2(\infty)$, and the capacitance will become $C_2(\infty) = \varepsilon_s/w_2(\infty)$, so that $C_1 > C_2(\infty) > C_2(0)$. The rate of increase of capacitance from $C_2(0)$ to $C_2(\infty)$ is determined by the time constant of the traps, and it has been shown by Senechal and Basinski (1968a) that the time variation obeys the equation

$$\frac{\Delta V_r}{\Delta C^{-2}} = \frac{q\varepsilon_s}{2}[N_d + N_t\{1-\exp(-t/\tau)\} - \lambda N_t C_{hm}\{1-\exp(-t/\tau)\}/\varepsilon_s] \tag{4.17}$$

where

$$\Delta V_r = V_2 - V_1$$

$$\Delta C^{-2} = \{C_2(t)\}^{-2} - C_1^{-2}$$

and C_{hm} is the harmonic mean $2C_1 C_2(t)/\{C_1 + C_2(t)\}$. Here τ is equal to e_e^{-1}, and N_d and N_t are assumed constant.

Eqn (4.17) can be understood in the following way. If we let the voltage increment ΔV become infinitesimally small so that $\Delta V/\Delta C^{-2}$ becomes $\partial V/\partial C^{-2}$ and C_{hm} becomes C, then eqn (4.17) becomes identical to eqn (4.16a) for $t = 0^+$. This is because the traps cannot instantaneously follow the change in bias, so that case (iii) of § 4.4.2 applies. If $t \to \infty$, eqn (4.17) becomes identical with eqn (4.15a) because the traps are able to respond to the change in bias after an infinitely long time and case (ii) applies. For arbitrary values of t, we may regard eqn (4.17) as derived from eqn (4.15a) if we suppose that a fraction $\{1-\exp(-t/\tau)\}$ of the traps have been able to follow the change in bias.

If the measurements made by observing $C_2(t)$ as a function of t for various values of V_2 are analysed by plotting $\Delta V_r/\Delta C^{-2}$ as a function of

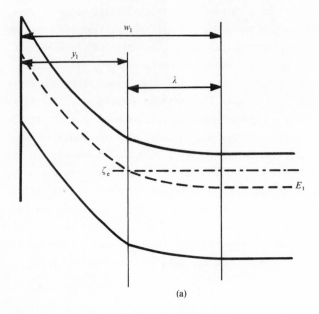

(a)

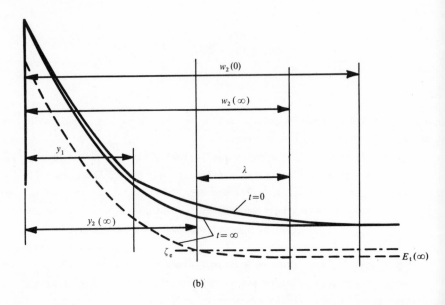

(b)

FIG. 4.11. Shape of Schottky barrier containing deep traps (a) with reverse bias V_1 (steady state), (b) after increase of reverse bias to V_2, showing shape of barrier at $t = 0$ and $t = \infty$.

C_{hm} for particular values of t, the relationship should be linear as
confirmed by Fig. 4.12 which shows Senechal and Basinski's experimental
results for Au–GaAs Schottky barriers. It can be seen that the straight
lines corresponding to different values of t all intersect at a common
point; this must occur when the right-hand side of eqn (4.17) is independent of t, i.e. when

$$\lambda C_{hm}/\varepsilon_s = 1. \tag{4.18}$$

The value of $\Delta v_r/\Delta c^{-2}$ at which the intersection occurs, namely $q\varepsilon_s N_d/2$, is
the same as would be obtained if t/τ were vanishingly small. The latter

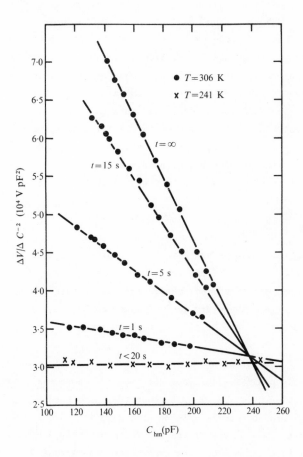

FIG. 4.12. Plots of $(\Delta v/\Delta c^{-2})$ against the harmonic mean capacitance C_{hm} for
a Au–GaAs Schottky barrier (Senechal and Basinski 1968b).

condition can be obtained by cooling the specimen so that the electron-
emission rate becomes very small and the trap-time constant very long. It
is seen from Fig. 4.12 that the data for T = 241 K do in fact give a constant value of $\Delta v_r/\Delta c^{-2}$ which passes through the intersection point
within experimental error.

By extrapolating the straight lines in Fig. 4.12 to cut the vertical
axis, one obtains $N_d + N_t\{1-\exp(-t/\tau)\}$ as a function of t, and from these
data N_d, N_t, and τ can be found. The value of N_d obtained in this way
agrees with the value obtained from the intersection point. Knowing the
value of C_{hm} at the intersection point, we may use eqn (4.18) to find λ
and hence obtain the energy of the trap with respect to the Fermi level in
the interior of the semiconductor from eqn (4.11). An alternative way of
obtaining an approximate value of the trap energy is to measure τ (and
hence e_e) as a function of temperature. Eqn (4.8a) can then be used to
find $E_t - E_i$ if the temperature variation of σ_e, v, and n_i can be ignored.

Lang (1974) has discussed a variant of the transient technique in which
the reverse bias is reduced for a short time and is then returned to its
original value. The capacitance decreases during the pulse, because more
traps become neutral, and then returns to its original value with a time
constant which is measured directly by means of a special dual-gated
detector. Traps with energy below E_1 have their population determined by
the hole quasi-Fermi level and give a capacitance transient of the opposite
sign.

4.4.4. The effect of light

Numerous investigations of traps have been reported which depend on the
change in the charge state of the traps brought about by exposure to light
and the subsequent effect of this change in charge state on the capacitance
of a Schottky barrier. There are many variants of the method, all of which
are generally described by the term photocapacitance.

Consider first the case of a Schottky barrier in an n-type semiconductor
containing donor-like traps which are all occupied by electrons and therefore electrically neutral. (This may come about because the traps all lie
below the zero-bias Fermi level or because they have been filled by an
application of forward bias as will be discussed in § 4.5.2.) Now suppose
that reverse bias is applied. If the trap population is determined by the
electron quasi-Fermi level, all the traps within a distance y of the metal
(Fig. 4.10(a)) will gradually lose their electrons by thermal emission.
However, at sufficiently low temperatures (say ~100 K), the thermal-

emission rate will be so low that the traps will retain their electrons
for a very long time. The capacitance measured at a moderately high frequency will be determined by the total positive space charge in the depletion region, which is the charge density qN_d due to the shallow donors. If
the barrier is now uniformly illuminated with light the quantum energy of
which exceeds the difference in energy between the trap and the bottom of
the conduction band, the traps will lose their electrons by photoexcitation
and become positively charged. Retrapping of the photoexcited electrons
can be neglected because they are removed as soon as they enter the conduction band by the electric field in the barrier, so the density of posi-
tively charged traps will tend towards N_t with a time constant $\tau_0 = (\varphi\sigma_0)^{-1}$
where φ is the flux of incident photons per unit area per second and σ_0 the
optical cross-section of the traps for electron emission. The square of the
capacitance will increase by a factor $(N_d+N_t)/N_d$ with the same time constant τ_0, so that if φ is known a measurement of τ_0 allows the optical
cross-section to be determined. The energy of the traps can be determined
from the photothreshold.

This technique was first used by Sah, Rosier, and Forbes (1969) using
p—n junctions. It is capable of great sensitivity because of the comparative ease with which small changes in capacitance can be measured. (A
change of one part in 10^5 is easily detectable if the initial capacitance
is around 10^{-10} F.) Sah and his co-workers were able to detect a concentration of 10^{17} m^{-3} sulphur atoms in silicon, which corresponded to a total
number in the depletion region of only 10^5, using this method. It has been
adapted to Schottky barriers by many workers, especially by Henry, Kukimoto,
Miller, and Merritt (1973) and by Hamilton (1974). A review has been given
by Grimmeiss (1974). A similar technique in which the traps are emptied by
raising the temperature (thermally stimulated capacitance) instead of by
photoexcitation has been used by Sah, Chan, Fu, and Walker (1972).

Williams (1966) has applied optical techniques to capacitance measurements in a different way. He applied reverse bias to gallium-arsenide
Schottky barriers at room temperature so that the traps in the depletion
region lost their electrons by thermal excitation. He then allowed light
of quantum energy greater than the semiconductor band gap to fall onto the
metal which was thin enough to transmit about 10% of the light. The light
generated hole—electron pairs in the gallium arsenide which increased the
reverse current. The electron concentration in the conduction band was
thereby increased to n', and the traps began to fill by electron capture
until finally a fraction α given by

$$\alpha = \frac{n'v'\sigma_e}{e_e + n'v'\sigma_e} = \frac{J_p\sigma_e}{J_p\sigma_e + qe_e}$$

was occupied by electrons, J_p being the photocurrent, σ_e the electron-capture cross-section of the trap, and v' the high-field saturation velocit of electrons in gallium arsenide. The corresponding time constant was $(e_e + n'v'\sigma_e)^{-1}$. It was therefore possible to determine both e_e and σ_e so that the trap energy could be calculated from eqn (4.8a), while the change in capacitance due to the application of reverse bias before illumination gave N_t.

A comprehensive analysis of almost all conceivable transient capacitance effects in p—n junctions subjected to all possible external stimuli has been given by Sah, Forbes, Rosier, and Tasch (1970). Most of their analysis is equally applicable to Schottky barriers. A comparison of the various methods has been given by Lang (1974).

4.5. THE CAPACITANCE UNDER FORWARD BIAS

4.5.1. *The diffusion capacitance*

A p—n junction under forward bias exhibits a capacitance due to the injection of minority carriers into the neutral regions on each side of the junction in addition to the space-charge capacitance of the depletion region. This diffusion capacitance increases exponentially with the forward-bias voltage and, for moderate values of forward bias, it dominates the space-charge capacitance. The diffusion capacitance is closely linked to the phenomenon of minority-carrier storage in which a p—n junction suddenly switched from forward to reverse bias exhibits a comparatively high reverse current until the injected minority carriers are cleared away. One can think of this transient current as discharging the diffusion capacitance.

It is natural to ask whether a Schottky diode may also exhibit diffusion capacitance when forward biased. We saw in § 3.7 that there is virtually no minority-carrier storage in a Schottky diode because the electrons injected into the metal rapidly lose energy through electron—electron collisions they cannot then negotiate the barrier if the bias is suddenly reversed. A corollary of this is that a Schottky diode exhibits no diffusion capacitance, and it is for this reason that when C/V measurements are used to find the height of a Schottky barrier they are frequently extended into forward bias, as the total capacitance is still determined by the capacitance of the depletion region. It is often difficult to make measurements

in forward bias, however, because of the large shunt conductance of the
diode.

4.5.2. *The effect of traps*

We shall now consider what determines the electron population of traps
under forward bias. Suppose there is a uniform distribution of donor-like
traps of energy E_t lying above E_1 so that $e_e > e_h$ (Fig. 4.13(a)). If ζ_e
lies slightly above E_t at the interface as shown in the diagram, $E_t + \zeta_e$
will exceed $2E_1$ throughout the depletion region so that $\sigma_e n\bar{v} > e_e$ because
of eqn (4.9b) and $\sigma_e n\bar{v} > e_h$ everywhere because of eqn (4.9c). We may
neglect e_h in the numerator and both e_e and e_h in the denominator of eqn
(4.7) and write

$$f \approx \frac{\sigma_e n\bar{v}}{\sigma_e n\bar{v} + \sigma_h p\bar{v}} \; .$$

To the right of the plane $x = b$ at which $\zeta_e + \zeta_h = 2E_1$, $\sigma_e n\bar{v} > \sigma_h p\bar{v}$ and
the traps are nearly all occupied. But for $x < b$, $\sigma_h p\bar{v} > \sigma_e n\bar{v}$ and f falls
below unity. However, if E_1 were a little lower than in the diagram (corresponding to a smaller value of σ_h/σ_e), b could approach zero and the
traps would then be filled right up to the surface of the metal. A similar
argument applied to traps lying below E_1 (Fig. 4.13(b)), for which $e_h > e_e$,
shows that to the left of the plane $x = b$ at which $\zeta_e + \zeta_h = 2E_1$ the trap
occupation is determined by the hole quasi-Fermi level so that the traps
are empty for $x < g$.

The assumption that is usually made is that the occupation of the traps
is everywhere determined by the electron quasi-Fermi level ζ_e; as we have
seen, it is not always true and great care must be taken in applying it.
It is more likely to be valid in large-band-gap semiconductors like gallium phosphide than in smaller-band-gap materials like silicon, because
E_1 lies much lower in the former case. It is also more likely to be valid
the nearer the trap lies to the conduction band.

Where the trap occupation is determined by the electron quasi-Fermi
level, the application of a forward bias such that ζ_e lies above the trap
level at the interface is often used to fill all the traps and so provide
a convenient set of initial conditions for transient capacitance or photocapacitance studies. It has also been used by Roberts and Crowell (1970,
1973) as the basis of a technique of measuring the energy level of the
traps. Their method relies on the fact that, if the capacitance of a
Schottky barrier containing traps is measured with a test frequency ω_s low

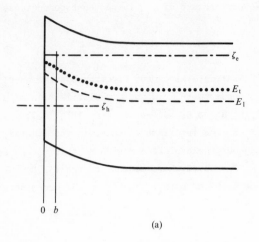

(a)

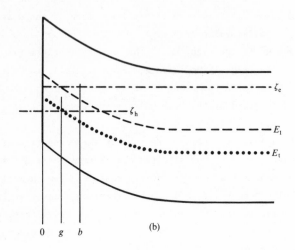

(b)

FIG. 4.13. Schottky barrier with deep traps under forward bias (a) $E_t > E_1$, (b) $E_t < E_1$.

enough for the traps to respond (case (i), § 4.4.2), the traps will make a contribution to the capacitance as long as ζ_e intersects the trap level E_t. If a forward bias is applied that is sufficiently large to make ζ_e lie above E_t at the interface, the traps will remain full throughout the period of the test frequency and will make no contribution to the capacitance. This condition reveals itself as a point of inflection in the plots of c^{-2} against v.

5. Practical contacts

In previous chapters, apart from admitting the possibility of an insulating interfacial layer, we have assumed that all contacts are metallurgically ideal in the sense that they involve an abrupt transition from a pure metal to a perfectly homogeneous semiconductor. In practice the methods of preparation are such that this ideal is never attained, and the purpose of this chapter is to discuss the deviations of practical contacts from ideally abrupt junctions.

5.1. METHODS OF MANUFACTURE

5.1.1. *Point contacts*

Although point contacts were extensively used as detectors in the early days of radio and later as microwave mixers and detectors, they have now almost completely given way to extended-area contacts because of the vastly superior reproducibility and reliability of the latter. The electrical properties of point contacts are very far from ideal and depend very much on the way the contacts are 'formed'. They have been extensively discussed in the books by Torrey and Whitmer (1948) and by Henisch (1957).

5.1.2. *Evaporated contacts*

The overwhelming majority of practical contacts are made by evaporation. Most of them are made in a conventional vacuum system pumped by a diffusion pump giving a vacuum around 10^{-5} Torr, often without a liquid-nitrogen trap. This method of depositing metal films has been extensively discussed by Holland (1956). The lower-melting-point metals such as aluminium and gold can usually be evaporated quite simply by resistive heating from a boat or filament, while the refractory metals like molybdenum and titanium are generally evaporated by electron-beam heating. Most frequently the semiconductor surface is prepared by chemical etching, and this invariably produces a thin oxide layer of thickness about 10–20 Å; the precise nature and thickness depend on the exact method of preparing the surface. The effect of surface preparation on the characteristics of silicon Schottky barriers has been discussed by Turner and Rhoderick (1968). Interfacial layers can also be caused by water or other vapour adsorbed onto the surface of the semiconductor before insertion into the vacuum system. Such adsorbed layers can usually be removed by heating the substrate to between 100°C and 200°C prior to evaporation (Smith 1969b).

Even if the semiconductor surface were devoid of an oxide layer to start with, as would be the case if it were prepared by cleaving a crystal so as to expose a fresh surface, an oxide layer would form by exposure to air in the time taken to transfer the semiconductor to a vacuum chamber and pump the system down to its final pressure. For example, in a vacuum of 10^{-5} Torr the gas molecules which strike the surface would, if they all adhered, build up a monolayer in about 10^{-1} s. If the sticking coefficient is as low as 10^{-3}, as it is believed to be for oxygen on silicon (Joyce and Neave 1971), it would take much longer (~100 s) for a monolayer to build up. However, the semiconductor surface is exposed to much higher pressures for several tens of minutes during the pump-down period, so there is plenty of time for an oxide to form.

The pump-down period can be avoided if the surface is prepared by cleavage or ion etching within the vacuum system after the final pressure has been reached, but an oxide layer can still be formed during the evaporation unless the vacuum is maintained at 10^{-7} Torr or better while the evaporation is in progress. To maintain pressures of this order requires a very good vacuum system, because the heat produced during the evaporation accelerates outgassing. An interfacial layer may also form as a result of the adsorption of vapour from a diffusion pump, and Cowley (1966) has shown that this source of contamination can be avoided by the use of an ion pump. For this reason it is becoming common practice to use ion-pumped systems capable of achieving a vacuum of 10^{-7} to 10^{-8} Torr.

The only way of ensuring that there is no interfacial layer is to evaporate a film onto a semiconductor immediately after it has been cleaved in an ultra-high vacuum at a pressure around 10^{-10} Torr (Thanailakis 1974). The time taken for a monolayer of adsorbed gas atoms to form at this pressure is so long (~10^7 s) that there is virtually no possibility of an interfacial layer being established or of the metal film being contaminated. But although contacts prepared in this way may be chemically ideal, they are still not physically ideal because of the mechanical damage to the surface during the cleaving process, and they usually show current/voltage characteristics with ideality factors n considerably greater than unity. This method of preparing contacts is far too slow and cumbersome to be used in industrial processes, and there is no potential advantage in doing so.

5.1.3. *Sputtered contacts*

Several authors have described contacts in which the metal has been deposited by sputtering. For example, Gutknecht and Strutt (1971) have reported aluminium—silicon diodes which were prepared by r.f. sputtering, though they give no details of either the sputtering gas or the pressure. They used 'sputter-etching' (i.e. sputtering in which the silicon was shielded from the aluminium) to etch the surface of the silicon prior to deposition of the metal. The diodes were almost ideal, with ideality factors of between 1.01 and 1.02 and barrier heights of 0.78 V, which coincides with the value obtained on cleaved surfaces (see § 2.8.1). Presumably the sputter-etching produces an almost ideally clean surface which has approximately the same surface-state density as a cleaved surface. Sinha and Poate (1973) have reported the fabrication of nearly ideal Schottky diodes of tungsten on gallium arsenide by r.f. sputtering.

Mullins and Brunnschweiler (1976) have made molybdenum—silicon Schottky diodes by d.c. sputtering in an argon atmosphere at a pressure of 0.1 Torr. Again they used sputter-etching to clean the surface of the silicon. They found that diodes prepared using a sputtering voltage of 1 kV were almost ideal, but that for greater sputtering voltages the I/V characteristics became non-ideal to an extent which depended on the voltage and on the sputtering time. They were able to explain their results by supposing that the sputtering produced defects near the surface of the silicon within a depth which increased with the d.c. voltage, the density of these defects being a linear function of the sputtering time. The defects apparently behaved like positively charged donors, and the additional space charge produced by these donors caused a narrowing of the barrier which in turn led to tunnelling of the electrons. They were able to give a good explanation of the observed I/V characteristics by assuming an exponential distribution of defects below the surface with a characteristic length in the range 10—100 Å. The almost ideal characteristics observed by Gutknecht and Strutt were presumably due either to the different sputtering conditions used by these workers or to the fact that they annealed their diodes after the sputtering had taken place.

Sputtering is frequently used to make contacts to practical devices because it produces metal films with good mechanical adhesion. It is also often used to deposit metal films prior to silicide formation (see § 5.3).

5.1.4. *Chemical deposition*

Surprisingly little attention has been given to the possibility of depositing the metal chemically, perhaps because of the comparative ease with
which the lower-melting-point metals can be evaporated. But because of
its simplicity and cheapness, chemical deposition is an attractive alternative especially for refractory metals. Crowell, Sarace, and Sze (1965)
deposited tungsten on germanium, silicon, and gallium arsenide by reacting
tungsten hexafluoride with the appropriate semiconductor. The resulting
Schottky diodes were close to ideal with n values not exceeding 1.04. Kano,
Inoue, Matsuno, and Takayanagi (1966) prepared molybdenum—silicon diodes
by reducing molybdenum pentachloride with hydrogen and found them to have
nearly ideal rectifying characteristics. Furakawa and Ishibashi (1967)
report having made ohmic and rectifying contacts to n-type and p-type
gallium arsenide, respectively, by reducing stannous chloride with hydrogen.

Deposition from solution has received comparatively little attention.
Gol'dberg, Posse, and Tsarenkov (1971) reported the manufacture of almost
ideal Schottky diodes from gallium phosphide by depositing gold or nickel
by an electroless process. For gold they used a mixture of $HAuCl_4$ and HF,
and for nickel a mixture of $NiCl_2$, NaH_2PO_2, $H_3C_6H_5O_7$, and NH_4Cl. The diodes
prepared in this way had n values of 1.02—1.03. The same authors have prepared nearly ideal Schottky diodes on gallium arsenide 'by chemical precipi-
tation of gold' but give no details of the process (Gol'dberg *et al.* 1975).
A list of solutions suitable for plating contacts onto gallium arsenide has
been given by Gol'dberg, Nasledov, and Tsarenkov (1971). Dörbeck (1966) has
reported the fabrication of rectifying contacts on gallium arsenide by
electroplating gold from an aqueous solution of gold and potassium cyanide
or copper from a solution of potassium copper cyanide. Both metals gave the
same barrier height (0.86 eV) which differs slightly from the values
usually obtained by evaporation.

Sullivan (1976) has described a novel method of depositing strongly
adherent films of silver, gold, palladium, and platinum from commercial
plating solutions onto gallium arsenide, gallium phosphide, and silicon.
The area of the (n-type) semiconductor on which deposition is required is
first damaged by bombardment with charged particles either from an accelera-
tor or from a glow discharge. The semiconductor is then immersed in the
plating solution and illuminated by light with sufficient quantum energy
to produce electron—hole pairs. Deposition of metal occurs on the damaged

area by an electrolytic process in which the damaged area acts as cathode and any undamaged area as anode, no external connections being necessary. Both ohmic and rectifying contacts were obtained; these depended on the nature of the bombarding particles, on the contact metal, and on post-fabrication annealing. Contacts which were not annealed were ohmic. Although the rectifying contacts obtained in this way are far from ideal, they seem to be adequate for many applications.

The information available in the literature about chemical-deposition methods is extremely sparse. This is probably because from a technological standpoint the methods may not be compatible with other processes involved in fabricating a semiconductor device, and also because those that are compatible tend to be shrouded in commercial secrecy. A review of the plating of metals on semiconductors has been given by Hillegas and Schnable (1963).

5.2. THE EFFECTS OF HEAT TREATMENT

Most contacts used in semiconductor devices are subjected to heat treatment. This may be deliberate, to promote adhesion of the metal to the semiconductor, or unavoidable, because high temperatures are needed for other processing stages which occur after the metal is deposited. Deliberat heating is often rather loosely referred to as 'sintering' or 'firing'. It is important to avoid the melting of rectifying contacts because the interface may become markedly non-planar with sharp metallic spikes projecting into the semiconductor. When this occurs, tunnelling through the high-field region at the tip of the spike may severely degrade the electrical characteristics (Andrews 1974). Unless alloying of the contact is desired (e.g. in the formation of ohmic contacts), it is necessary to keep the temperatures to which contacts are subjected below the eutectic temperature of the metal—semiconductor system. For example, the eutectic temperatures of alloys of silicon with the three common contact metals gold, aluminium, and silver are 370°C, 577°C, and 840°C, respectively.

Even at temperatures substantially below the eutectic temperature, migration of the semiconductor through the metal may occur. These metallurgical changes have been extensively studied recently by using the newly developed techniques of Rutherford backscattering, Auger electron spectroscopy, and secondary-ion mass-spectrometry. (These techniques are described in review articles by Mayer and Turos (1973) and by Morgan (1975).) For example, Hiraki, Lugujjo, Nicolet, and Mayer (1971) have shown by using Rutherford backscattering that at temperatures as low as 200°C there can be substantial migration of silicon through a gold film. The extent of the

migration is very sensitive to the condition of the surface of the silicon
before evaporation of the gold, and can be almost completely suppressed by
the presence of a thin oxide film at the interface. It cannot be explained
simply in terms of homogeneous diffusion but probably involves grain-
boundary diffusion as well. The effect of the migration is to make the
interface depart very considerably from a perfect silicon—metal junction,
and the electrical characteristics become very non-ideal. Generally, speaking, the diffusion constant of silicon in a metal is many orders of magnitude greater than the diffusion constant of the same metal in silicon, so
that diffusion on the silicon side of the interface can usually be ignored
(McCaldin 1974).

It is usually difficult to relate the degradation of the I/V characteristics to the observed metallurgical changes. The change in the I/V characteristics cannot normally be explained simply in terms of a change in barrier height, but as a rule the whole shape of the characteristic alters to
such an extent that it is clear that we no longer have a simple Schottky
barrier. Occasionally the characteristic can be explained by supposing that
atoms which behave as donors or acceptors diffuse into the semiconductor,
or that electrically active defects are created so that the effective density of dopant in the semiconductor is changed. If the dopant density increases, the barrier gets thinner and thermionic-field emission (or even
field emission) may occur. If the diffusing atoms or defects introduced
into the semiconductor have the opposite polarity to that of the original
dopant, the effective dopant density decreases and it occasionally happens
that a p—n junction may form. A good example of this is the case of
aluminium—silicon contacts which have been extensively studied because of
their technological importance. Chino (1973) first reported that aluminium—
silicon contacts heated above 450°C showed a significant change in I/V
characteristics which, in the case of n-type silicon, could be described in
terms of increases in the barrier height and ideality factor. This change
in characteristics was accompanied by pronounced pitting of the silicon
surface. Basterfield, Shannon, and Gill (1975) have given a convincing
explanation of these observations in the following terms. At temperatures
around 500°C silicon is taken up into solid solution by the aluminium to an
amount determined by the solubility limit at the particular temperature.
On cooling the dissolved silicon recrystallizes onto the n-type silicon as
an aluminium-doped p-type layer, because aluminium is an acceptor. The concentration of aluminium in the recrystallized silicon at 500°C is about
5×10^{24} m^{-3}. The net space-charge density is therefore negative near the

interface, and the bands become bent downwards as shown in Fig. 5.1. If the distance from the maximum of the barrier to the interface is less than the electron mean free path, the structure behaves like a Schottky barrier of height φ'_b.

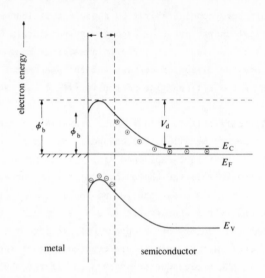

FIG. 5.1. Band diagram for aluminium–silicon contact after heat treatment (Basterfield, Shannon, and Gill 1975).

The aluminium concentration quoted above is of the right order of magnitude to cause an increase in barrier height of the order of 0.1 eV which is what was observed in practice. This model has been extended by Card (1974), and Card and Singer (1975) have shown by using Auger electron spectroscopy that, if aluminium–silicon contacts are heated to 500°C for 20 min, the concentration profiles of the various elements are changed as shown in Fig. 5.2. In the heat-treated sample the aluminium signal continues a further 100–200 Å into the silicon than in the untreated sample. The aluminium concentration in this tail is of the order of 10^{25} m^{-3}; this is adequate to explain the change in the electrical characteristics. If the silicon is p-type, the aluminium tail causes the depletion region to become narrower, and the effective barrier height is reduced as a result of tunnelling (Card 1975).

Another semiconductor which has been extensively investigated because of its technological importance is gallium arsenide. Contacts to gallium

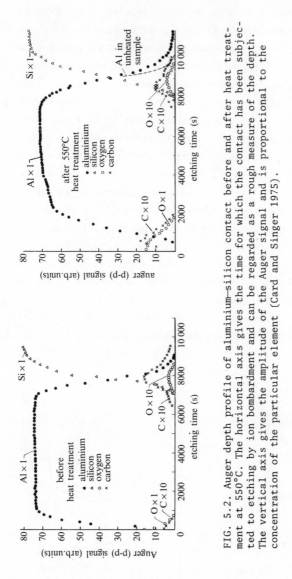

FIG. 5.2. Auger depth profile of aluminium–silicon contact before and after heat treatment at 550°C. The horizontal axis gives the time for which the contact has been subjected to etching by ion bombardment and can be regarded as a rough measure of the depth. The vertical axis gives the amplitude of the Auger signal and is proportional to the concentration of the particular element (Card and Singer 1975).

arsenide have been studied by Sinha and Poate (1973) using Rutherford scattering; by Todd, Ashwell, Speight, and Heckingbottom (1974) using Rutherford scattering and Auger electron spectroscopy; and by Kim, Sweeney, and Heng (1974) using secondary-ion mass-spectrometry. The results of their investigations are that tungsten and aluminium contacts are stable up to 500°C, whereas gold and platinum contacts show considerable degradation in their electrical characteristics due to extensive metallurgical changes at

the interface. Fig. 5.3 shows the I/V characteristics of gold–gallium-
arsenide diodes obtained by Kim *et al.* after heat treatment to various
temperatures, and Fig. 5.4 shows the corresponding gold and gallium distributions. Above 350°C there is extensive migration of gallium into the gold
up to a limiting concentration set by the solid solubility, followed by a
rapid increase in the gallium concentration at the Au–GaAs eutectic temperature of 450°C. At this temperature there is also substantial migration
of gold into the gallium arsenide. The arsenic (not shown) remains relatively stationary up to 450°C but diffuses rapidly into the gold as the
temperature approaches 500°C. The diffusion of gallium into the gold is
accompanied by the formation of gallium vacancies in the gallium arsenide.
Madams, Morgan, and Howes (1975) have suggested that these vacancies act as
donors and that the change in the I/V characteristics arises not because of
a change in barrier height but because the increased donor density causes
thermionic-field emission. Because the concentration of gallium in the gold
is determined by its solid solubility at temperatures below the eutectic
temperature, the total number of gallium atoms present in the gold must be
proportional to the thickness of the gold film, so that a thin film will
have a less serious effect on the characteristics of the diode than a thick
one.

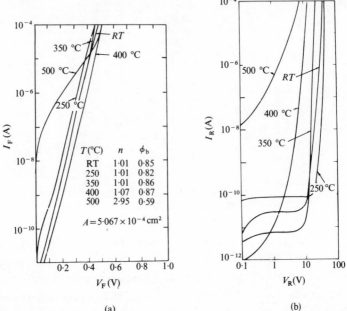

$T(°C)$	n	ϕ_b
RT	1·01	0·85
250	1·01	0·82
350	1·01	0·86
400	1·07	0·87
500	2·95	0·59

$A = 5·067 \times 10^{-4} \, \text{cm}^2$

(a)   (b)

FIG. 5.3. (a) Forward and (b) reverse characteristics of an Au–GaAs contact
after heat treatment (Kim, Sweeney, and Heng 1974. Copyright Institute of
Physics).

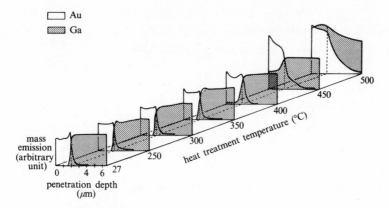

FIG. 5.4. Secondary-ion mass-spectra of a Au—GaAs contact after heat treatment (Kim, Sweeney, and Heng 1974. Copyright Institute of Physics).

In device applications thermal degradation may come about as a result of
heating caused by the passage of current through the diode. Gerzon, Barnes,
Waite, and Northrop (1975) have examined the 'burn-out' mechanism in metal—
silicon microwave mixers subjected to pulses of microwave power. They found
a close correlation between the power necessary to cause burn-out and the
temperature at which burn-out can be simulated by heating alone. This temperature (and therefore the amount of power necessary to cause burn-out)
was found to depend on the choice of metal.

5.3. SILICIDES

The effect of heat on metal—silicon contacts is particularly important if
the metal is capable of forming a silicide which is a stoichiometric compound. Most metals, including all the transition metals, form silicides
after appropriate heat treatment as Table 5.1 shows. These silicides may
form as a result of solid-state reactions at temperatures of about one-
third to one-half the melting point of the silicide in degrees K (Mayer
and Tu 1974). The vast majority of silicides exhibit metallic conductivity
(Neshpor and Samsonov 1960) so that, if a metallic silicide is formed as a
result of heat treatment of a metal—silicon contact, the silicide—silicon
junction behaves like a metal—semiconductor contact and may exhibit rectifying properties. Moreover, because the silicide—silicon interface is
formed some distance below the original surface of the silicon it is free
from contamination and very stable at room temperature; contacts formed in
this way generally show stable electrical characteristics which are very

close to ideal. They also exhibit very good mechanical adhesion. Silicide–silicon Schottky-barrier diodes were first described by Lepselter and Sze (1968) and by Andrews and Lepselter (1969, 1970).

TABLE 5.1.

Silicide formation in periodic system (McCaldin 1974)

Li	Be													
Na	Mg											Al		
K	Ca	Sc	Ti	V	Cr	Mn	Fe	Co	Ni	Cu	Zn	Ga	Ge	
Rb	Sr	Y	Zr	Nb	Mo		Ru	Rh	Pd	Ag	Cd	In	Sn	Sb
Cs	Ba	La	Hf	Ta	W	Re	Os	Ir	Pt	Au	Hg	Tl	Pb	Bi

All metals form silicides except those enclosed in boxes.

5.3.1. Reaction kinetics

The kinetics of silicide formation have been extensively studied using the techniques of Auger electron spectroscopy, X-ray diffraction, and especially Rutherford scattering. Fig. 5.5 shows the Rutherford backscattering of 1.75 MeV He^+ ions by a 1500 Å film of rhodium deposited on silicon (Coe, Rhoderick, Gerzon, and Tinsley 1974). The Figure shows the energy distribution of the He^+ ions scattered through an angle of 160° after heat treatment of the contact for 20 min at various temperatures. The results corresponding to 377°C (curve A) are indistinguishable from those obtained with no heat treatment. The group of He^+ ions with energies between 1.15 and 1.50 MeV are those scattered by the nuclei of rhodium atoms in the metal film, the ions of higher energy being scattered from the outer surface of the rhodium film and the ions of lower energy from the surface of the film adjacent to the silicon. The loss of energy of these ions is due partly to the loss of energy of the incident ions through interaction with electrons in penetrating the film, and partly to the recoil energy given up to the rhodium nuclei. The He^+ ions with energies below 0.75 MeV are those scattered by silicon nuclei. These have a much lower energy than those scattered by the rhodium nuclei. This is partly because the incident ions contributing to this group have passed through the whole of the rhodium film, but mainly because the recoil energy given up to a silicon nucleus is much greater than that given up to a rhodium nucleus as a result of the much smaller atomic weight of silicon.

At 483°C (curve B) conversion of rhodium to RhSi has begun. The reaction begins at the rhodium—silicon interface and moves outwards through the metal. The ions with energies between 1.10 and 1.30 MeV are those scattered by rhodium nuclei in the RhSi phase. The yield (counts per unit

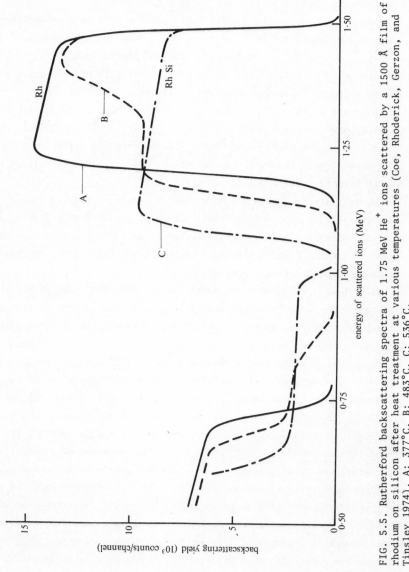

FIG. 5.5. Rutherford backscattering spectra of 1.75 MeV He$^+$ ions scattered by a 1500 Å film of rhodium on silicon after heat treatment at various temperatures (Coe, Rhoderick, Gerzon, and Tinsley 1974). A: 377°C, B: 483°C, C: 536°C.

energy range) of these ions has decreased because the concentration of
rhodium atoms in RhSi is less than that in rhodium metal. In addition the
minimum energy of the group of ions scattered by rhodium nuclei has been
reduced because the energy lost by the incident ions in penetrating both
the unconverted rhodium and the RhSi is greater than the energy lost in the
original rhodium film. At the same time the maximum energy of the group of
ions scattered by silicon nuclei has increased to about 0.90 MeV because
silicon atoms have moved into the rhodium to form RhSi.

Finally, at 536°C (curve C) the conversion to RhSi is complete. The
yield of ions scattered by rhodium nuclei is constant as a result of the
uniform concentration of rhodium in RhSi, and the maximum energy of the
ions scattered by silicon has increased to a value (1.00 MeV) corresponding
to scattering by silicon nuclei located at the outer surface of the RhSi.
The thickness of the RhSi phase was found to increase linearly with time;
this indicates that the rate of growth is limited by the reaction rate at
one or both of the phase boundaries rather than by the diffusion of silicon
or rhodium through the RhSi. Similar studies of the platinum—silicon system
have been carried out by Poate and Tisone (1974).

Another silicide system which has been extensively studied because of
its technological importance is the nickel—silicon system (Andrews and Koch
1971; Tu, Alessandrini, Chu, Kraütle, and Mayer 1974; Coe 1974; Coe and
Rhoderick 1976). Three silicide phases have been identified. The first to
form at a temperature of approximately 300°C is Ni_2Si. This is well established within a period of a few hours. After about ten hours at the same
temperature the NiSi phase appears. Both phases appear first at the silicon
surface and move progressively towards the outer surface of the nickel. The
thickness of the Ni_2Si phase is proportional to the square root of the
reaction time, whereas that of the NiSi phase is linearly proportional to
the time. This implies that the rate of growth of the Ni_2Si phase is limited
by the diffusion of one of the elements, whereas that of the NiSi phase is
limited by the reaction rate at either or both of the phase boundaries.
Above 750°C a third phase, $NiSi_2$, appears. Unlike the two earlier reactions,
the conversion to $NiSi_2$ is effected, not by the movement of a planar interface between adjacent phases, but by a gradual decrease in the average
nickel concentration throughout the NiSi over a temperature range of about
100°C. This suggests that there is nucleation of the di-silicide at sites
throughout the NiSi phase rather than at the phase boundaries. Similar behaviour has been reported by Ziegler, Mayer, Kircher, and Tu (1973) to
occur in the conversion of HfSi to $HfSi_2$. Such a process is likely to occur
if grain growth is important. The Rutherford-scattering spectra indicate

that the $NiSi_2$—Si interface is far from abrupt, and scanning electron microscopy shows pronounced pitting of the surface (Coe 1974), so it is clear that we are no longer dealing with a simple 1-dimensional problem. An additional feature of the growth of $NiSi_2$ is that it is epitaxial on the (111), (110), and (100) surfaces of silicon. This probably reflects the very close match in crystal structure and lattice spacing between the disilicide and silicon.

Pitting of the surface can often result from heat treatment, particularly if there is any trace of oxide on the surface of the silicon before the metal is deposited. A very thin oxide film can severely inhibit silicide formation, and if the oxide is not uniform there may be uneven growth of the silicide due to this cause. It is important to realise that the appearance of a smeared-out phase transition in Rutherford-scattering spectra may be due to a macroscopically non-planar phase boundary rather than to a gradual phase transition, and it is important to check the flatness of the surface with a scanning electron microscope before any attempt is made to interpret the experimental data.

5.3.2. *Barrier heights*

The Schottky diodes which result from silicide formation usually show nearly ideal characteristics with *n*-values of around 1.1 or less, and with good agreement between the values of the barrier height determined from the *I/V* and *C/V* characteristics. Table 5.2 shows barrier heights associated with the rhodium—silicon and nickel—silicon systems (Coe 1974). The most obvious feature of these results is the remarkable constancy of the barrier height, except for the two highest temperatures in the rhodium case. If one regards the silicide—silicon junction as forming an ideal Schottky barrier to which the Bardeen theory can be applied, the constancy of the barrier height implies either that the work functions of the metal and the silicides are remarkably similar, or else that the silicon surface is behaving like a cleaved surface and that the barrier height is clamped by surface states. There is almost no information available about work functions of silicides, and it is difficult to believe that the silicide—silicon interface is so abrupt that it has the properties of a cleaved surface. A much simpler explanation, but one which is difficult to test experimentally, is that there may be a very thin layer of one particular phase always present immediately adjacent to the silicon, and that this layer is thin enough to escape detection by Rutherford scattering or other techniques but thick enough to determine the barrier height. A layer of silicide

TABLE 5.2.

Barrier heights of Ni–Si and Rh–Si systems (Coe 1974)

Reaction temperature (°C)	Phases present	Barrier height (eV)
20	Ni–Si	0.70
230	Ni–(Ni$_2$Si)–Si	0.71
324	Ni–Ni$_2$Si–(NiSi)–Si	0.70
356	Ni$_2$Si–NiSi–Si	0.70
377	Ni$_2$Si–NiSi–Si	0.72
430	NiSi–Si	0.69
550	NiSi–Si	0.68
650	NiSi–(NiSi$_2$)–Si	0.69
850	NiSi$_2$–Si	0.70
20	Rh–Si	0.78
324	Rh–Si	0.78
377	Rh–(RhSi)–Si	0.79
430	Rh–RhSi–Si	0.79
536	RhSi–Si	0.74
600	RhSi–Si	0.70

about 50 Å thick would have properties sufficiently close to those of the bulk solid to control the barrier height, but would not be resolved by any of the analytical techniques currently available. Bearing in mind that most evaporation sources entail some radiative heating of the substrate, that the deposition of the metal results in a release of condensation energy (especially if sputtering is involved), and that silicide formation can take place at comparatively low temperatures, it seems quite likely that a thin layer of silicide may exist even in the absence of any heat treatment. After heat treatment, one would certainly not expect any of the phase boundaries to be perfectly abrupt like ideal metal–semiconductor interfaces.

5.4. CONTROL OF BARRIER HEIGHTS

Practical applications often arise in which it is desirable to be able to control the barrier height of a rectifying contact. One simple method which suggests itself is the use of an alloy of two metals, as was reported by Arizumi, Hirose, and Altaf (1968, 1969) who used alloys of the noble metals. They were able to obtain a linear variation of barrier height with composition, but the practical application of this principle would require extremely careful control of the composition of the alloy. Another method which is possible in principle is to control the thickness or properties of the interfacial layer. Archer and Atalla (1963) and Saltich and Terry (1970) have discussed the effect of exposure of the semiconductor surface to oxygen before evaporation of the metal on the barrier heights of silicon Schottky diodes. However, it has not yet been found possible to apply this method in a controllable manner to make stable diodes.

The only practicable method so far reported seems to be that of Shannon (1976), who has shown that the effective height of a Schottky barrier can be controlled over quite a wide range by incorporating highly doped surface layers. These highly doped layers were realized in practice by using ion implantation. To reduce the effective barrier height, the surface layer was doped with impurities of the same type as those in the interior of the semiconductor (i.e. donors in the case of n-type semiconductors). The increase in the doping level reduced the thickness of the depletion region to such an extent that thermionic-field emission took place, with a resulting increase in the saturation current density which is equivalent to a decrease in barrier height. By careful control of the donor concentration and distribution in the surface layer so that the depletion region just filled the heavily doped layer under zero bias, it was found possible to maintain good reverse characteristics. The extra field produced by the application of reverse bias was generated predominantly within the low-doped bulk of the semiconductor, so that the surface field was not altered much and there was comparatively little increase in the thermionic-field-emission current. In this way Shannon was able to reduce the height of nickel—silicon Schottky barriers (φ_b = 0.58 eV) by up to 0.2 eV by implanting 5 keV antimony ions with a mean range of about 50 Å and a surface concentration of 10^{17} m^{-2}. The lower curve in Fig. 5.6 shows the effective barrier height for nickel on n-type silicon as a function of the surface concentration of antimony ions.

To increase the effective barrier height, Shannon used surface layers doped with impurities of the opposite type to those in the interior of the semiconductor (i.e. acceptors in the case of n-type semiconductors). The effect of the implanted acceptors is similar to that of the aluminium atoms in n-type silicon described in § 5.2. If the concentration is great enough, the bands will actually bend downwards as shown in Fig. 5.1 and, if the distance of the maximum of the barrier from the metal is less than the

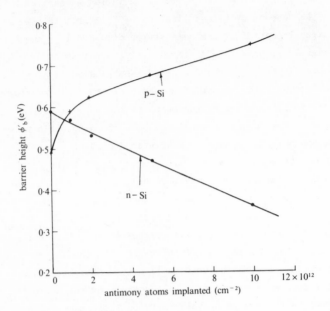

FIG. 5.6. Effective barrier height of nickel contacts on n-type and p-type
silicon after implantation with various doses of 5 keV antimony ions
(Shannon 1974).

electron mean free path, the system behaves like a Schottky barrier of
height φ_b'. Shannon used this principle to increase the height of the
Schottky barrier produced by nickel on p-type silicon by implanting antimony ions. The upper curve in Fig. 5.6 shows the effective barrier height
as a function of the surface concentration of 5 keV antimony ions. The
same method has been used by Ponpon and Siffert (1975) to increase the
height of barriers on n-type silicon by implanting gallium ions.

5.5. OHMIC CONTACTS

An ohmic contact is one for which the I/V characteristic is determined by
the resistivity of the semiconductor specimen or by the behaviour of the
device of which the contact forms part, rather than by the characteristics
of the contact. It is not essential that the I/V characteristic of the contact itself is linear, provided its resistance is very small compared with
the resistance of the specimen or device. In addition, the contact should
not inject minority carriers and should be stable both electrically and
mechanically.

An important property of an ohmic contact is its specific resistance

(resistance multiplied by area). A good ohmic contact should have a specific resistance of less than about 10^{-7} Ω m^2. The specific resistance R_c can
be found by measuring the total resistance of a circular contact of dia-
meter d on a slice of semiconductor of thickness t and resistivity ρ. The
total resistance R_{tot} is given (Cox and Strack 1967) by

$$R_{tot} = \frac{\rho}{\pi d} \tan^{-1}\left(\frac{4t}{d}\right) + \frac{4R_c}{\pi d^2} + R_0$$

where R_0 is the resistance of the back contact.

The fabrication of ohmic contacts is still more of an art than a science,
and every laboratory tends to have its own favourite recipes which involve
particular metal or alloy systems, particular deposition methods, and particular forms of heat treatment. A comprehensive list of recipes for ohmic
contacts to germanium, silicon, III–V and II–VI compounds has been given by
Milnes and Feucht (1972), and to III–V compounds by Rideout (1975).

All the recipes appear to depend on one or other of the following three
principles:

(a) If the semiconductor is one which conforms approximately to the
    simple Mott theory (eqn (2.2)), it should be possible to create an
    ohmic contact by finding a metal with a work function less than the
    work function of an n-type semiconductor or greater than the work
    function of a p-type semiconductor, as discussed in § 2.2.1. Unfor-
    tunately there are very few metal–semiconductor combinations with
    this property. If the inequality is nearly but not quite satisfied,
    the result should be a rectifying contact with a very low barrier
    height which in practice may serve effectively as an ohmic contact.

(b) The vast majority of ohmic contacts depend on the principle of
    having a thin layer of very heavily doped semiconductor immediately
    adjacent to the metal. The depletion region is then so thin that
    field emission takes place and the contact has a very low resistance
    at zero bias. The mechanism has already been discussed in § 3.3.2.

(c) If the surface of the semiconductor is damaged (for example, by
    sand-blasting), crystal defects may be formed near the surface which
    act as efficient recombination centres. If the density of these is
    high enough, recombination in the depletion region will become the
    dominant conduction mechanism and will cause a significant decrease
    in the contact resistance.

The most widely used of these methods is the one described in (b). The
heavily doped layer may be formed separately from the deposition of the

metal, say by diffusion or by ion implantation (Meyer 1969), or it may
result from the deposition and subsequent heat treatment of an alloy containing an element which acts as a donor or acceptor in the semiconductor.
On heating, the semiconductor dissolves in the metal, and on subsequent
cooling it crystallizes out with a high concentration of the electrically
active element in solid solution. For silicon and germanium, the alloys
most commonly used are gold–antimony for n-type and gold–indium for p-type.
For gallium arsenide, which has received much attention as far as ohmic
contacts are concerned because of its microwave applications, the most
common alloys are those of gold or silver with germanium or tin (for n-type
material) or with manganese (for p-type material). One of the most important properties of a suitable alloy for ohmic contacts is that it should
'wet' the surface of the semiconductor when it melts (Zettlemoyer 1969).
For instance, gold–germanium does not wet the surface of gallium arsenide
very well, and tends to 'ball up' when heated. This can be overcome by
depositing a thin film of nickel on top of the gold–germanium which has
the effect of keeping the latter continuous. A metallurgical study of the
Ni/Au–Ge/GaAs system has been made by Robinson (1975). Sometimes alloys
will form equally good ohmic contacts on either p- or n-type material;
for example, gold–germanium will work on both p- and n-type gallium
arsenide because germanium is an amphoteric impurity (i.e. it can behave
as either a donor or as an acceptor) in gallium arsenide.

   A method of making ohmic contacts which has the merit of great simplicity but is only applicable in a limited number of cases is that of plating
by deposition from a solution without the passage of an electric current
('electroless' plating). To ensure that the metal should adhere well it is
often found to be necessary to roughen the surface of the semiconductor.
Sullivan and Eigler (1957) made ohmic contacts to silicon by the electroless plating of nickel from a solution of nickel chloride, sodium hypophosphite, ammonium citrate, and ammonium chloride. The process was carried
out at 100°C. On low-resistivity material ($\approx 5 \times 10^{-4}$ Ω m) the method gave
low-resistance contacts to p-type material provided the contact was not
annealed and a low resistance to n-type material after annealing to 600°C.
The p-type result is presumably due to the existence of a low barrier, the
n-type result to the penetration of phosphorus from the hypophosphite into
the silicon at elevated temperature.

Appendix A: The depletion approximation

It is frequently necessary to know how the electrostatic potential and electric-field strength in a Schottky barrier depend on the barrier height, bias voltage, and impurity concentration, and for most purposes it is sufficiently accurate to use an approximation known as the depletion approximation. In this approximation the free-carrier density is assumed to fall abruptly from a value equal to the density in the bulk of the semiconductor to a value which is negligible compared with the donor or acceptor concentration, whereas in reality the transition occurs smoothly over the distance in which the bands bend by about $3kT/q$.

Let us consider the case of an n-type semiconductor. Outside the depletion region the semiconductor is neutral and within the depletion region the charge density is equal to qN_d, so that the variation of charge density with distance is as shown in Fig. A.1. With no applied bias there can be no electric field within the neutral region of the semiconductor otherwise a current would flow; even when a bias voltage is applied, the current is usually so small in relation to the conductivity of the semiconductor that the electric field in the neutral region is negligible. The electric field is related to the charge density by Gauss's theorem so that $\partial\mathscr{E}/\partial x = qN_d/\varepsilon_s$. The magnitude of $\mathscr{E}$ (which is negative since it is directed in the negative x direction*) increases linearly as the metal is approached, rising from zero at $x = w$ to a value of $qN_d w/\varepsilon_s$ at the interface. If we take the electrostatic potential ψ to be zero within the neutral region so that $\psi(w) = 0$, the potential at x is given by

$$\psi(x) = \int_x^w \mathscr{E}dx$$

$$= -\int_x^w \frac{qN_x(w-x)}{\varepsilon_s}\,dx$$

$$= -\frac{qN_d}{2\varepsilon_s}(w-x)^2.$$

---

*This definition is necessary to enable us to write Gauss's theorem in the above form, which implies that the positive direction of $\mathscr{E}$ must be in the positive x direction. However, in Chapter 2 it was convenient to use the opposite definition of the sign of $\mathscr{E}$ so that the field would be positive in an n-type semiconductor.

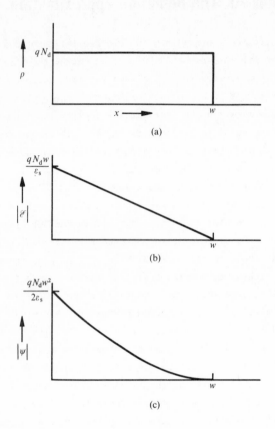

FIG. A.1. Variation of (a) charge density, (b) electric-field strength, and (c) electrostatic potential with distance according to the depletion approximation.

The magnitude of ψ rises quadratically as the metal is approached and has the value $qN_d w^2/2\varepsilon_s$ at the interface. This result can be remembered easily because the difference in potential across the depletion region is simply equal to the average electric-field strength multiplied by w, and if $\mathscr{E}$ varies linearly the average field strength is equal to $\frac{1}{2}|\mathscr{E}_{max}| = \frac{1}{2}qN_d w/\varepsilon_s$. The magnitude of ψ at the interface is equal to the diffusion potential V_d so that

$$V_d = \frac{qN_d w^2}{2\varepsilon_s}. \tag{A.1}$$

Making use of the relationships

$$\mathscr{E}_{max} = \frac{qN_d w}{\varepsilon_s} = \frac{Q_d}{\varepsilon_s} ,$$

where Q_d is the total charge per unit area due to the uncompensated donors in the depletion region, we can write eqn (A.1) in the alternative forms

$$V_d = \frac{\varepsilon_s |\mathscr{E}_{max}|^2}{2qN_d} \tag{A.2}$$

and

$$V_d = \frac{Q_d^2}{2\varepsilon_s qN_d} . \tag{A.3}$$

According to eqn (A.3) the differential capacitance per unit area is given by

$$C = \frac{\partial Q_d}{\partial V_d} = \left(\frac{\varepsilon_s qN_d}{2V_d}\right)^{\frac{1}{2}} = \frac{\varepsilon_s}{w} . \tag{A.4}$$

The energy of the bottom of the conduction band *relative to the Fermi level in the metal* (eqn (3.6)) is given by

$$E_c(x) = \varphi_b + \{\psi(0) - \psi(x)\}$$

$$= \varphi_b + \frac{qN_d}{2\varepsilon_s}(x^2 - 2wx) . \tag{A.5}$$

Appendix B: Exact analysis of the electric field in a Schottky barrier

It is sometimes necessary to know more accurately than the depletion approximation allows how the electric-field strength in a Schottky barrier depends on the barrier height, bias voltage, and impurity concentration. The analysis will be developed for an n-type semiconductor.

Let the electrostatic potential within the semiconductor at a distance x from the interface be $\psi(x)$, and suppose that the zero of potential is chosen so that $\psi(\infty) = 0$. The energy of the bottom of the conduction band at a point in the depletion region then exceeds that in the interior of the semiconductor by an amount $-q\psi$, and if the bands are bent upwards as in Fig. B.1 ψ must be negative. Let the equilibrium density of electrons in the interior of the semiconductor be n_0 and the equilibrium density of holes immediately adjacent to the interface be p_s. At a distance x from the interface the densities of electrons and holes depend on the positions of their respective quasi-Fermi levels (see § 3.2). For electrons the thermionic-emission theory assumes that the quasi-Fermi level remains horizontal throughout the depletion region and coincides with the Fermi level in the bulk semiconductor and, even according to the diffusion theory, the quasi-Fermi level is still flat almost up to the interface. Assuming the semiconductor to be non-degenerate, the electron concentration in the depletion region is given by $n = n_0 \exp(q\psi/kT)$ where n_0 is the electron concentration in the neutral region of the semiconductor (see Fig. B.1). For holes the usual assumption (which has been justified by Rhoderick 1972) is that the hole quasi-Fermi level is horizontal throughout the depletion region and coincides with the Fermi level in the metal, so the hole concentration is given by

$$p = p_s \exp[-q(V_d + \psi)/kT]$$

where V_d is the diffusion potential (or band bending) and p_s is the hole concentration at $x = 0$. V_d is equal to $-\psi_s$ where ψ_s is the surface potential.

The net charge density is given by

$$\rho = q(N_d + p - n)$$

$$= q[N_d + p_s \exp\{-q(V_d + \psi)/kT\} - n_0 \exp(q\psi/kT)].$$

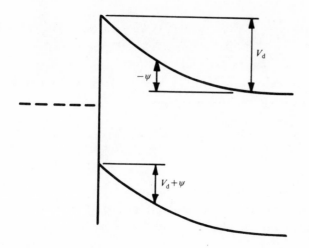

FIG. B.1. Band diagram of a Schottky barrier, showing definition of ψ and V_d.

The charge density and electrostatic potential are related by Poisson's equation

$$\frac{d^2\psi}{dx^2} = - \frac{\rho}{\varepsilon_s} .$$

It is convenient to introduce new variables $u = q\psi/kT$ and $F = -du/dx$ so that $d^2u/dx^2 = F(dF/du)$. Hence

$$F\frac{dF}{du} = - \frac{q^2}{\varepsilon_s kT} \{N_d + p_s \exp(u_s - u) - n_0 \exp(u)\}$$

where u_s ($= -qV_d/kT$) is the value of u at the surface. Integrating from the interior of the semiconductor to the surface,

$$F_s^2 = - \frac{2q^2}{\varepsilon_s kT} \{N_d u_s - p_s (1 - \exp(u_s)) - n_0 (\exp(u_s) - 1)\}$$

where F_s is the value of F at the surface. We have made use of the fact that F and u both vanish in the interior of the semiconductor and N_d is assumed constant. The electric-field strength is given by

$$\mathcal{E} = -d\psi/dx = -(kT/q)(du/dx) = kTF/q$$

so that

$$\mathscr{E}^2_{max} = - \frac{2kT}{\varepsilon_s}\{N_d u_s - p_s(1-\exp(u_s)) - n_0(\exp(u_s)-1)\}.$$

Writing this equation in terms of V_d $(= -kTu_s/q)$ and making use of the fact that for an n-type semiconductor n_0 is very closely equal to N_d (assuming complete ionization of the donors), we find

$$\mathscr{E}^2_{max} = \frac{2q}{\varepsilon_s}\left\{N_d\left(V_d - \frac{kT}{q}\right) + \frac{p_s kT}{q}(1-\exp(-qV_d/kT)) + \frac{kTN_d}{q}\exp(-qV_d/kT)\right\}, \qquad (B.1)$$

an equation first derived by Atalla (1966).

Unless the barrier height is so large that the surface is strongly p-type, the term involving p_s is negligible, and if $qV_d \gg kT$ we may neglect the exponential term in the coefficient of N_d so that, to a very good approximation,

$$\mathscr{E}^2_{max} = \frac{2qN_d}{\varepsilon_s}\left(V_d - \frac{kT}{q}\right). \qquad (B.2)$$

The first term on the right can be understood very simply because it is the result obtained from the depletion approximation in which the charge density is assumed to rise abruptly from zero to the value qN_d at the edge of the depletion region. The second term is a correction due to the transition region (sometimes incorrectly referred to as the 'reserve' layer*) in which the electron density falls from n_0 to a value negligible compared with N_d.

---

*The 'reserve' regime of a semiconductor refers to the situation in which the donors are not completely ionized. That is not the case here.

Appendix C: Comparison of Schottky diodes and p-n junctions

Although the choice between Schottky diodes and p—n junctions for device applications is usually dictated by purely technological considerations, it sometimes happens that both of them are technologically feasible, and the choice between them has then to be made by reference to their electrical properties. A full discussion of this topic has been given by Rhoderick (1976).

C.1. CURRENT-TRANSPORT MECHANISMS

We shall compare a Schottky diode made from an n-type semiconductor with a p—n junction having the same barrier height. (By the 'barrier height' of a p—n junction we mean the quantity φ_b in Fig. C.1.) The Schottky diode has the property that the current is carried almost entirely by electrons even though the semiconductor may be only moderately doped, and to ensure that this is also true of the p—n junction we must assume that the latter is made with the p side more lightly doped than the n side.

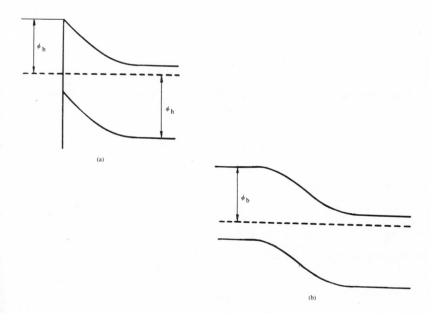

FIG. C.1. Band diagram of (a) Schottky barrier and (b) p—n junction, showing definition of φ_b and φ_h.

Assuming that the current through the p–n junction is carried entirely by electrons, the current/voltage relationship will be of the form

$$J_{pn} = J_{0pn}\{\exp(qV/kT)-1\}$$ (C.1)

where

$$J_{0pn} = \frac{qD_e N_c}{L_e} \exp(-q\varphi_b/kT).$$ (C.2)

Here N_c is the effective density of states in the conduction band and D_e and L_e are the diffusion constant and diffusion length, respectively, of electrons in the p-type side. Eqn (C.2) can be rewritten by making use of the relationship (from kinetic theory) that $D_e = \bar{v} l_e/3$, where l_e is the mean free path of electrons in the p region and $\bar{v}$ their mean thermal velocity. Moreover $l_e = \bar{v} \tau_{ce}$ and $L_e = \sqrt{D_e \tau_{re}}$, where τ_{ce} is the mean time between collisions and τ_{re} the lifetime of electrons in the p-type side; eqn (C.2) then becomes

$$J_{0pn} = \{qN_c\bar{v}/(3r_e)^{\frac{1}{2}}\}\exp(-q\varphi_b/kT)$$ (C.3)

where

$$r_e = \tau_{re}/\tau_{ce}.$$

For a Schottky diode the current/voltage relationship is normally given by the thermionic-emission theory, according to which

$$J_S = J_{0S}\{\exp(qV/kT)-1\}.$$ (C.4)

From eqn (3.9) we may write

$$J_{0S} = (qN_c\bar{v}/4)\exp(-q\varphi_b/kT).$$ (C.5)

Comparison of eqns (C.3) and (C.5) gives

$$\frac{J_{0S}}{J_{0pn}} = \frac{(3r_e)^{\frac{1}{2}}}{4} \approx r_e^{\frac{1}{2}}.$$ (C.6)

For a silicon p–n junction, a typical value of τ_{re} is $\sim 10^{-6}$ s while

$\tau_{ce} \sim 10^{-13}$ s, so $(r_e)^{\frac{1}{2}} > 10^3$. Hence for the same barrier height the saturation-current density of a Schottky diode exceeds that of a silicon p—n
junction by a factor of 10^3 or more. For a short-lifetime semiconductor
like gallium arsenide the ratio is smaller but is still of the order of
10^2.

   In practice, the barrier height of a Schottky diode is likely to be
appreciably less than that of a p—n junction made from the same semiconductor. In the case of silicon, for example, the lowest possible value of φ_b,
which occurs when the p side of the p—n junction is very lightly doped (say
with $N_a \approx 10^{20}$ m^{-3}), is about 0.8 eV, while it is easy to make Schottky
diodes with barrier heights in the range 0.5 to 0.6 eV. This fact, taken in
conjunction with the argument of the previous paragraph, has the result
that the saturation current density of a Schottky diode is likely to exceed
that of a p—n junction made from the same semiconductor by a factor of 10^7
or more. Expressed in a different way, for the same current density the
forward voltage drop across a p—n junction will exceed that across a
Schottky diode by at least 0.4 V. This makes a Schottky diode particularly
suitable for use as a low-voltage high-current rectifier. Conversely, since
the reverse saturation current is also larger, a Schottky diode is not as
suitable as a p—n junction for use as a high-voltage low-current rectifier.

C.2. HOLE INJECTION

In the case of a p—n junction, the term 'minority-carrier injection' refers
to the injection of those carriers which are predominantly responsible for
current flow. For the case under consideration this means the injection of
electrons into the p-type side, and it is this sort of minority-carrier
injection which is utilized in n—p—n transistors and light-emitting diodes.
The exact analogue of this process for a Schottky diode made from an n-type
semiconductor consists of the injection of electrons into the metal. These
injected electrons do not influence the conductivity of the metal in any
way, so there are no effects attributable to the direct analogue of
minority-carrier injection in a Schottky diode.

   However, the injection of holes from the metal into the semiconductor,
although usually very small, is not always completely negligible. The analogue of this process in the p—n junction under consideration would be the
injection of holes into the n-type side; this is generally negligible if
the n side is much more heavily doped than the p side. The transport of
holes through the depletion region of a Schottky barrier and their subsequent diffusion in the neutral region of the semiconductor is identical to

the mechanism of hole transport in a p–n junction, and we may write

$$J_h = J_{0h}\{\exp(qV/kT)-1\},$$

where, by analogy with eqn (C.3),

$$J_{0h} = \{qN_v\bar{v}/(3r_h)^{\frac{1}{2}}\}\exp(-q\varphi_h/kT). \tag{C.7}$$

Here N_v is the effective density of states in the valence band, φ_h is the 'barrier for holes' shown in Fig. C.1, and $r_h = \tau_{rh}/\tau_{ch}$. The hole injection-ratio γ_h is therefore given by

$$\gamma_h = \frac{J_h}{J_e + J_h} \approx \frac{J_h}{J_e} = \frac{4N_v}{(3r_h)^{\frac{1}{2}}N_c} \exp\{-q(\varphi_h-\varphi_b)/kT\}, \tag{C.8}$$

or

$$\gamma_h \approx \frac{1}{(r_h)^{\frac{1}{2}}} \exp\{-q(\varphi_h-\varphi_b)/kT\}, \tag{C.9}$$

since J_e is the thermionic-emission current given by eqns (C.4) and (C.5) and $N_c \approx N_v$. Eqn (C.8) is equivalent to eqn (3.30) if the concentration gradient of holes is determined by the hole diffusion length L_h rather than by the thickness of the quasi-neutral region L. As eqn (C.9) shows, γ_h increases with increasing φ_b because of the reduction in J_e and decreases with increasing φ_h because of the reduction in J_h. For most Schottky barriers $\varphi_h > \varphi_b$ so the exponent is negative. Taking the case of a gold–silicon Schottky barrier with $\varphi_b = 0.8$ eV and $N_d = 10^{22}$ m^{-3} and assuming $(r_h)^{\frac{1}{2}} \approx 10^3$, we find $\gamma_h \approx 10^{-5}$; this is in reasonable agreement with Yu and Snow's results shown in Fig. 3.12. This doping level corresponds to $\varphi_h \approx 0.9$ eV. If we take $N_d = 10^{20}$ m^{-3}, φ_h is reduced to 0.8 eV and γ_h increases to $\approx 10^{-3}$. The hole injection-ratio is small in a p–n junction (assuming it to be asymmetrically doped) but is even smaller for a Schottky diode because, although the hole current should be the same for a given donor density in both cases, the electron current is much greater in the Schottky diode because of the considerations of § C.1. From the point of view of potential applications, it is the *electron* injection-ratio which is of interest in the p–n junction under consideration (with $N_d > N_a$), and this is vastly greater than the hole-injection ratio in the Schottky diode.

As was mentioned in § 3.8, the hole injection-ratio of a Schottky diode can
be increased by the presence of a thin interfacial layer to an extent which
may have possible device applications.

One consequence of minority-carrier injection in a p–n junction is that
majority carriers are drawn in from the contact to maintain almost perfect
charge neutrality ('quasi-neutrality'). Since minority-carrier injection
refers to the carriers which emanate from the more strongly doped side, the
effect is to increase the carrier concentration in the more weakly doped
side. This effect is known as conductivity modulation and is of considerable importance in power rectifiers. In a Schottky diode the injection of
electrons into the metal has a negligible effect on the metal's conductivity. However, if the barrier height φ_b is large and the semiconductor
is of high resistivity (which gives a comparatively small value of φ_n),
the injection of holes into the semiconductor may cause observable conductivity modulation in the semiconductor at high current densities where γ_h
increases with J_e as discussed in § 3.5. (J_e is effectively equal to the
total current density.) Conductivity modulation of this sort has been observed by Jäger and Kosak (1973) in Schottky diodes made from high-
resistivity silicon.

C.3. MINORITY-CARRIER STORAGE

Minority-carrier storage is a consequence of minority-carrier injection and
in a p–n junction refers to the injection of those carriers predominantly
responsible for current flow, which for the case under consideration are
electrons. If the bias is suddenly changed from forward to reverse, the
electrons injected into the p side must be removed before the diode attains
a high-resistance state. The recovery time is approximately equal to τ_{re}
which in practice may be about 10^{-6} s for a silicon rectifier. The use of
p–n junctions as fast switching diodes or microwave mixers is limited by
this recovery time.

The exact analogue of minority-carrier storage for a Schottky diode
made from an n-type semiconductor is the storage of electrons after they
have been injected into the metal. As we saw in § 3.7, the time taken for
these injected electrons to lose so much energy that they are unable to
diffuse back into the semiconductor is about 10^{-14} s, which is quite negligible. There is therefore no direct analogue of minority-carrier storage in
a Schottky diode.

There is, however, the possibility of storage effects associated with
the injection of holes into the semiconductor. The recombination time of

the holes ($\sim 10^{-6}$ s) is quite long compared with the storage time associated with the electrons injected into the metal, but the total charge Q_h stored in this manner is approximately equal to $J_h \tau_{rh} = \gamma_h J_e \tau_{rh}$; this is still quite small because of the smallness of the hole injection-ratio γ_h. The effective storage time τ_h is given by $\tau_h = Q_h / J_{tot} \approx \gamma_h \tau_{rh}$ and is generally less than 10^{-11} s. (We have assumed that the total current density J_{tot} is effectively equal to J_e.) At high current densities τ_h rises because of the increase in γ_h and depends in a complicated way on the boundary conditions at the ohmic contact as has been discussed in some detail by Scharfetter (1965). Even τ_h is not observable experimentally, and in practice the recovery times of Schottky diodes are limited by their *RC* product.

Minority-carrier storage in a p–n junction has associated with it a capacitance — the diffusion capacitance — which is added to the ordinary depletion-region capacitance when the junction is forward biased. One can think of the charge stored in the injected carriers as constituting the charge on a capacitor. Because there is virtually no minority-carrier storage in Schottky diodes, there is no diffusion capacitance either, and the capacitance of a forward-biased Schottky barrier is simply that due to the space charge in the depletion region.

References

Allen, F.G. and Gobeli, G. W. (1962). *Phys. Rev.*, **127**, 150.
Anderson, C.L., Crowell, C.R., and Kao, T.W. (1975). *Solid-St. Electron.*, **18**, 705.
Andersson, L.P., Hyder, A., and Berg, S. (1973). *Nucl. Instrum. Meth.*, **114**, 1.
Andrews, J.M. (1974). *J. Vac. Sci. Technol.*, **11**, 972.
—— and Koch, F.B. (1971). *Solid-St. Electron.*, **14**, 901.
—— and Lepselter, M.P. (1969). *Ohmic contacts to semiconductors*, p. 159. Electrochem. Soc., New York.
—— —— (1970). *Solid-St. Electron.*, **13**, 1011.
—— and Phillips, J.C. (1975). *Phys. Rev. Lett.*, **35**, 56.
Archer, R.J. (1962). *J. Opt. Soc. Am.*, **52**, 970.
—— and Atalla, M.M. (1963). *Ann. N.Y. Acad. Sci.*, **101**, 697.
—— and Yep, T.O. (1970). *J. Appl. Phys.*, **41**, 303.
Arizumi, T. and Hirose, M. (1969). *Jap. J. Appl. Phys.*, **8**, 749.
—— —— and Altaf, N. (1968). *Jap. J. Appl. Phys.*, **7**, 870.
—— —— —— (1969). *Jap. J. Appl. Phys.*, **8**, 1310.
Arthur, J.R. (1966). *J. Appl. Phys.*, **36**, 3212.
Assimos, J.A. and Trivich, D. (1973). *J. Appl. Phys.*, **44**, 1687.
Atalla, M.M. (1966). *Proc. Munich Symp. Microelectron.*, p. 123. Oldenberg, Munich.
Baccarani, G. and Mazzone, A.M. (1976). *Electron. Lett.*, **12**, 59.
Baker, W.D. and Milnes, A.G. (1972). *J. Appl. Phys.*, **43**, 5152.
Bardeen, J. (1947). *Phys. Rev.*, **71**, 717.
—— and Brattain, W.H. (1949). *Phys. Rev.*, **75**, 1208.
Basterfield, J., Shannon, J.M., and Gill, A. (1975). *Solid-St. Electron.*, **18**, 290.
Baxandall, P.J., Colliver, D.J., and Fray, A.F. (1971). *J. Phys. E.: Sci. Inst.*, **4**, 213.
Bennett, A.J. and Duke, C.B. (1967). *Phys. Rev.*, **162**, 578.
Berz, F. (1974). *Solid-St. Electron.*, **17**, 1245.
Bethe, H.A. (1942). M.I.T. Radiation Lab. Rep. 43-12.
Blakemore, J.S. (1962). *Semiconductor statistics*. Pergamon Press, Oxford.
Braun, F. (1874). *Pogg. Ann.*, **153**, 556.
Braun, I. and Henisch, H.K. (1966). *Solid-St. Electron.*, **9**, 981.
Broom, R.F. (1971). *Solid-St. Electron.*, **14**, 1087.
Card, H.C. (1971). *Proc. Manchester Conf. Metal–Semicond. Contacts*, p. 129. Inst. of Phys., London.
—— (1975). *Solid State Commun.*, **16**, 87.
—— and Rhoderick, E.H. (1971a). *J. Phys. D: Appl. Phys.*, **4**, 1589.
—— —— (1971b). *J. Phys. D: Appl. Phys.*, **4**, 1602.
—— —— (1973). *Solid-St. Electron.*, **16**, 365.
—— and Singer, K.E. (1975). *Thin Solid Films*, **28**, 265.
Chang, C.Y. and Sze, S.M. (1970). *Solid-St. Electron.*, **13**, 727.
Chino, K. (1973). *Solid-St. Electron.*, **16**, 119.
Clarke, R.A., Green, M.A., and Shewchun, J. (1974). *J. Appl. Phys.*, **45**, 1442.
Coe, D.J. (1971). Paper 10.3 ESSDERC Conference, Munich.
—— (1974). Ph.D. Thesis, Manchester University.
—— and Rhoderick, E.H. (1976). *J. Phys. D: Appl. Phys.*, **9**, 965.
—— —— Gerzon, P.H., and Tinsley, A.W. (1974). *Proc. Manchester Conf. Metal– Semicond. Contacts*, p. 74. Inst. of Phys., London.

Consigny, R.L. and Madigan, J.R. (1969). *Solid State Commun.*, **7**, 189.
Copeland, J.A. (1969). *IEEE Trans. Electron Devices*, **ED-16**, 445.
Cowley, A.M. (1966). *J. Appl. Phys.*, **37**, 3024.
—— (1970). *Solid-St. Electron.*, **13**, 403.
—— and Sze, S.M. (1965). *J. Appl. Phys.*, **36**, 3212.
—— and Zettler, R.A. (1968). *IEEE Trans. Electron Devices*, **ED-15**, 761.
Cox, R.H. and Strack, H. (1967). *Solid-St. Electron.*, **10**, 1213.
Crowell, C.R. (1965). *Solid-St. Electron.*, **8**, 395.
—— (1977). *Solid-St. Electron.*, **20**, 171.
—— and Beguwala, M. (1971). *Solid-St. Electron.*, **14**, 1149.
—— and Rideout, V.L. (1969). *Solid-St. Electron.*, **12**, 89.
—— and Roberts, G.I. (1969). *J. Appl. Phys.*, **40**, 3726.
—— Sarace, J.C., and Sze, S.M. (1965). *Trans. Metall. Soc. AIME*, **233**, 478.
—— Shore, H.B., and La Bate, E.E. (1965). *J. Appl. Phys.*, **36**, 3843.
—— and Sze, S.M. (1966). *Solid-St. Electron.*, **9**, 1035.
—— —— and Spitzer, W.G. (1964). *Appl. Phys. Lett.*, **4**, 91.
Davydov, B. (1939). *J. Phys. USSR*, **1**, 167.
—— (1941). *J. Phys. USSR*, **4**, 335.
Deal, B.E., Snow, E.H., and Mead, C.A. (1968). *J. Phys. Chem. Solids*, **27**, 1873.
Dörbeck, F.H. (1966). *Solid-St. Electron.*, **9**, 1135.
Eastman, D.E. and Freeouf, J.L. (1975). *Phys. Rev. Lett.*, **34**, 1624.
Eimers, G.W. and Stevens, E.H. (1971). *IEEE Trans. Electron Devices*, **ED-18**, 1185.
Fowler, R.H. (1931). *Phys. Rev.*, **38**, 45.
Frankl, D.R. (1967). *Electrical properties of semiconductor surfaces*, p. 33. Pergamon Press, Oxford.
Furukawa, Y. and Ishibashi, Y. (1967). *Jap. J. Appl. Phys.*, **6**, 787.
Gatos, H.C., Moody, P.L., and Lavine, M.C. (1960). *J. Appl. Phys.*, **31**, 212.
Gerzon, P.H., Barnes, J.W., Waite, D.W., and Northrop, D.C. (1975). *Solid-St. Electron.*, **18**, 343.
Glover, G.H. (1973). *Solid-St. Electron.*, **16**, 973.
Gol'dberg, Y.A., Nasledov, D.N., and Tsarenkov, B.V. (1971). *Instrum. Exper. Tech.*, **3**, 899.
—— Posse, E.A., and Tsarenkov, B.V. (1971). *Electron. Lett.*, **7**, 601.
—— —— —— (1975). *Soviet Phys. Semicond.*, **9**, 337.
Goodman, A.M. (1963). *J. Appl. Phys.*, **34**, 329.
—— (1964). *J. Appl. Phys.*, **35**, 573.
—— and Perkins, D.M. (1964). *J. Appl. Phys.*, **35**, 3351.
Gossick, B.R. (1963). *Solid-St. Electron.*, **6**, 445.
Gray, K.E. (1973). *Solid State Commun.*, **13**, 1787.
Green, M.A. (1976). *Solid-St. Electron.*, **19**, 421.
—— and Shewchun, J. (1973). *Solid-St. Electron.*, **16**, 1141.
Gregory, P.E. and Spicer, W.E. (1975). *Phys. Rev.*, **12**, 2370.
Grimmeiss, H.G. (1974). *Proc. Manchester Conf. Metal—Semicond. Contacts*, p. 187. Inst. Phys., London.
Grondahl, L.O. (1926). *Phys. Rev.*, **27**, 813.
—— (1933). *Rev. Mod. Phys.*, **5**, 141.
Gutknecht, P. and Strutt, M.J.O. (1971). *Electron. Lett.*, **7**, 298.
—— —— (1972). *Appl. Phys. Lett.*, **21**, 405.
—— —— (1974). *IEEE Trans. Electron Devices*, **ED-21**, 172.
Hackam, R. and Harrop, P. (1972). *Solid State Commun.*, **11**, 669.
Haeri, S.Y. and Rhoderick, E.H. (1974). *Proc. Manchester Conf. Metal— Semicond. Contacts*, p. 84. Inst. Phys., London.
Hall, R.N. (1952). *Phys. Rev.*, **87**, 387.
Hamilton, B. (1974). *Proc. Manchester Conf. Metal—Semicond. Contacts*, p. 218. Inst. Phys., London.
Heine, V. (1965). *Phys. Rev.*, **138**, A 1689.

Heine, V. (1972). *Proc. Roy. Soc.* A, **331**, 307.

Henisch, H.K. (1957). *Rectifying semiconductor contacts*. Oxford University Press, Oxford.

Henry, C.H., Kukimoto, H., Miller, G.L., and Merritt, F.R. (1973). *Phys. Rev.*, **B7**, 2499.

Hillegas, W.J. and Schnable, G.L. (1963). *Electrochem. Tech.*, **1**, 228.

Hiraki, A., Lugujjo, E., Nicolet, M.-A., and Mayer, J.W. (1971). *Phys. Status Solidi*, **A7**, 401.

Hirose, M., Altaf, N., and Arizumi, T. (1970). *Jap. J. Appl. Phys.*, **9**, 260.

Holland, L. (1956). *Vacuum Deposition of Thin Films*. Chapman & Hall, London.

Honeisen, B. and Mead, C.A. (1971). *Solid-St. Electron.*, **14**, 1225.

Hooper, R.C., Cunningham, J.A., and Harper, J.G. (1965). *Solid-St. Electron.*, **8**, 831.

Inkson, J.C. (1973). *J. Phys. C. (Sol. State)*, **6**, 1359.

—— (1974). *J. Vacuum Sci. Technol.*, **11**, 943.

Jäger, H. and Kosak, W. (1969). *Solid-St. Electron.*, **12**, 511.

—— —— (1973). *Solid-St. Electron.*, **16**, 357.

James, H.M. (1949). *Phys. Rev.*, **76**, 1602.

Joffe, J. (1945). *Elect. Commun.*, **22**, 217.

Jones, R.O. (1969). *Structure and chemistry of solid surfaces*, pp. 14—1 to 14—15. Wiley, New York.

Joyce, B.A. and Neave, J.H. (1971). *Surf. Sci.*, **27**, 499.

Kahng, D. (1963). *Solid-St. Electron.*, **10**, 45.

—— (1964). *Bell Syst. Tech. J.*, **42**, 215.

Kajiyama, K., Sakata, S., and Ochi, O. (1975). *J. Appl. Phys.*, **46**, 3221.

Kano, G., Inoue, M., Matsuno, J., and Takayanagi, S. (1966). *J. Appl. Phys.*, **37**, 2985.

Kar, S. (1975). *Solid-St. Electron.*, **18**, 169.

Kennedy, D.P., Murley, P.C., and Kleinfelder, W. (1968). *IBM J. Res. Developm.*, **12**, 399.

Kim, H.B., Sweeney, G.G., and Heng, T.M.S. (1974). *Proc. 5th Int. Symp. Gallium Arsenide*, p. 307. Inst. Phys., London.

Kimerling, L.C. (1974). *J. Appl. Phys.*, **45**, 1839.

Korwin-Pawlowski, M.L. and Heasell, E.L. (1975). *Solid-St. Electron.*, **18**, 849.

Kumar, R.C. (1970). *Internat. J. Electron.*, **29**, 365.

Kurtin, S., McGill, T.C., and Mead, C.A. (1969). *Phys. Rev. Lett.*, **22**, 1433.

Kusaka, M., Matsui, T., and Okazaki, S. (1974). *Surf. Sci.*, **41**, 607.

Landkammer, F.J. (1967). *Solid State Commun.*, **5**, 247.

Lang, D.V. (1974). *J. Appl. Phys.*, **45**, 3023.

Lanyon, H.P.D. and Richardson, R.E. (1971). *Phys. Status Solidi*, A 7, 411.

Lepselter, M.P. and Sze, S.M. (1968). *Bell Syst. Tech. J.*, **47**, 195.

Levine, J.D. (1971). *J. Appl. Phys.*, **42**, 3991.

Livingstone, A.W., Turvey, K., and Allen, J.W. (1973). *Solid-St. Electron.*, **16**, 351.

Lodge, O. (1890). *JIEE*, **19**, 346.

Loeb, L.B. (1961). *The kinetic theory of gases*, p. 42. Dover Publications Inc., New York.

Losee, D.L. (1975). *J. Appl. Phys.*, **46**, 2204.

Louie, S.G. and Cohen, M.L. (1975). *Phys. Rev. Lett.*, **35**, 866.

Loveluck, J.M. and Rhoderick, E.H. (1967). *Solid-St. Electron.*, **10**, 433.

Madams, C.J., Morgan, D.V., and Howes, M.J. (1975). *Electron.Lett.*, **11**, 574.

Many, A., Goldstein, Y., and Grover, N.B. (1965). *Semiconductor surfaces*. North-Holland, Amsterdam.

Mayer, J.W. and Tu, K.N. (1974). *J. Vacuum Sci. Technol.*, **11**, 86.

—— and Turos, A. (1973). *Thin Solid Films*, **19**, 1.

McCaldin, J.O. (1974). *J. Vacuum Sci. Technol.*, **11**, 990.

McColl, M., Millea, M.F., and Mead, C.A. (1971). *Solid-St. Electron.*, **14**, 677.

Mead, C.A. (1965). *Appl. Phys. Lett.*, **6**, 103.
— (1966). *Solid-St. Electron.*, **9**, 1023.
— and McGill, T.C. (1976). *Phys. Lett.*, **58A**, 249.
— and Spitzer, W.G. (1964). *Phys. Rev.*, **134**, A 713.
Meyer, D.E. (1969). *Ohmic contacts to semiconductors*, p. 227. Electrochemical Society, New York.
Miller, W.L. and Gordon, A.R. (1931). *J. Phys. Chem.*, **35**, 2785.
Milnes, A.G. and Feucht, D.L. (1972). *Heterojunctions and metal—semiconductor junctions*. Academic Press, New York.
Morgan, D.V. (1975). *Contemporary Phys.*, **16**, 221.
Mott, N.F. (1938). *Proc. Cambridge Phil. Soc.*, **34**, 568.
— and Jones, H. (1936). *Theory of the properties of metals and alloys*. Oxford University Press, Oxford.
Mottram, J.D. (1977). Ph.D. Thesis, University of Manchester.
Mullins, F.H. and Brunnschweiler, A. (1976). *Solid-St. Electron.*, **19**, 47.
Neshpor, V.S. and Samsonov, G.V. (1960). *Soviet Phys. Solid State*, 2, 1966.
Neville, R.C. and Mead, C.A. (1970). *J. Appl. Phys.*, **41**, 3795.
Nguyen, P.H., Lepley, B., Nadeau, A., and Ravelet, S. (1975). *Proc. IEE*, **122**, 1193.
Nill, K.W., Walpole, J.N., Calawa, A.R., and Harman, T.C. (1970). *Proc. Conf. on Phys. of Semimetals and Narrow-Gap Semicond.*, p.383. Pergamon.
Okuto, Y. and Crowell, C.R. (1974). *J. Jap. Soc. Appl. Phys.*, **43** (Supp.), 390.
Padovani, F.A. (1967). *J. Appl. Phys.*, **38**, 891.
— (1968). *Solid-St. Electron.*, **11**, 193.
— (1971). *Semiconductors and semimetals* (ed. Willardson & Beer), 6A, Chap. 2. Academic Press, New York.
— and Stratton, R. (1966). *Solid-St. Electron.*, **9**, 695.
— and Sumner, G.G. (1965). *J. Appl. Phys.*, **36**, 3744.
Parker, G.H., McGill, T.C., Mead, C.A., and Hoffman, D. (1968). *Solid-St. Electron.*, **11**, 201.
Pauling, L. (1960). *The nature of the chemical bond*. Cornell University Press, Ithaca.
Pellegrini, B. (1973). *Phys. Rev.*, **7B**, 5299.
— (1974). *Solid-St. Electron.*, **17**, 217.
Perlman, S.S. (1969). *IEEE Trans. Electron Devices*, **ED-16**, 450.
Persky, G. (1972). *Solid-St. Electron.*, **15**, 1345.
Pickard, G.W. (1906). US Patent No. 836531.
Pierce, G.W. (1907). *Phys. Rev.*, **25**, 31.
Poate, J.M. and Tisone, T.C. (1974). *Appl. Phys. Lett.*, **24**, 391.
Ponpon, J.P. and Siffert, P. (1975). *J. de. Phys.*, **36**, L-149.
Pugh, D. (1964). *Phys. Rev. Lett.*, **12**, 390.
Rhoderick, E.H. (1970). *J. Phys. D: Appl. Phys.*, **3**, 1153.
— (1972). *J. Phys. D: Appl. Phys.*, **5**, 1920.
— (1974). *Proc. Manchester Conf. Metal—Semicond. Contacts*, p. 3. Inst. Phys., London.
— (1975). *J. Appl. Phys.*, **46**, 2809.
— (1976). *Proc. 5th European Solid-State Device Res. Conf.*, p. 103. Soc. Française de Physique, Paris.
Rideout, V.L. (1975). *Solid-St. Electron.*, **18**, 541.
— and Crowell, C.R. (1970). *Solid-St. Electron.*, **13**, 993.
Ritchie, R.H. and Ashley, J.C. (1965). *J. Phys. Chem. Solids*, 26, 1689.
Rivière, J.C. (1957). *Proc. Phys. Soc.*, **70**, 676.
— (1966). *Appl. Phys. Lett.*, **8**, 172.
— (1969). *Solid-State Surface Sci.*, 1, 179.
Roberts, G.I. and Crowell, C.R. (1970). *J. Appl. Phys.*, **41**, 1767.
— — (1973). *Solid-St. Electron.*, **16**, 29.
Robinson, G.Y. (1975). *Solid-St. Electron.*, **18**, 331.

Rowe, J.E., Christman, S.B., and Margaritondo, G. (1975). *Phys. Rev. Lett.*, **35**, 1471.

Ryder, E.J. and Shockley, W. (1951). *Phys. Rev.*, **81**, 139.

Sah, C.T., Chan, W.W., Fu, H.S., and Walker, J.W. (1972). *Appl. Phys. Lett.*, **20**, 193.

―― Forbes, L., Rosier, L.L., and Tasch, A.F. (1970). *Solid-St. Electron.*, **13**, 759.

―― Noyce, R.N., and Shockley, W. (1957). *Proc. IRE*, **45**, 1228.

―― Rosier, L.L., and Forbes, L. (1969). *Appl. Phys. Lett.*, **15**, 316.

Saltich, J.L. (1969). *Ohmic contacts to semiconductors*, p. 187. Electro-chemical Society, New York.

―― and Clark, L.E. (1970). *Solid-St. Electron.*, **13**, 857.

―― and Terry, L.E. (1970). *Proc. IEEE*, **58**, 492.

Saxena, A.N. (1969). *Surf. Sci.*, **13**, 151.

Scharfetter, D.L. (1965). *Solid-St. Electron.*, **8**, 299.

Schottky, W. (1938). *Naturwiss.*, **26**, 843.

―― (1939). *Z. Phys.*, **113**, 367.

―― (1942). *Z. Phys.*, **118**, 539.

―― and Spenke, E. (1939). *Wiss. Veroff. Siemens-Werken*, **18**, 225.

―― Störmer, R., and Waibel, F. (1931). *Z. Hochfrequenztechnik*, **37**, 162.

Schultz, W. (1954). *Z. Phys.*, **138**, 598.

Schwarz, R.F. and Walsh, J.F. (1953). *Proc. IRE*, **41**, 1715.

Sebenne, C., Bolmont, D., Guichar, G., and Balkanski, M. (1974). *Proc. 12th Int. Conf. Phys. Semicond.*, p. 1313. Teubner, Stuttgart.

Seiranyan, G.B. and Tkhorik, Y.A. (1972). *Phys. Status Solidi*, A **13**, K115.

Seitz, F. (1940). *Modern theory of solids*. McGraw-Hill, New York.

Senechal, R.R. and Basinski, J. (1968a). *J. Appl. Phys.*, **39**, 3723.

―― and Basinski, J. (1968b). *J. Appl. Phys.*, **39**, 4581.

Shannon, J.M. (1974). *Appl. Phys. Lett.*, **25**, 75.

―― (1976). *Solid-St. Electron.*, **19**, 537.

Shockley, W. (1950). *Electrons and holes in semiconductors*. Van Nostrand, New York.

―― and Read, W.T. (1952). *Phys. Rev.*, **87**, 835.

Sigurd, D. (1974). *Proc. Manchester Conf. Metal—Semicond. Contacts*, p. 141. Inst. Phys., London.

Sinha, A.K. and Poate, J.M. (1973). *Appl. Phys. Lett.*, **23**, 666.

Smith, B.L. (1968). *Electron. Lett.*, **4**, 332.

―― (1969a). Ph.D. Thesis, Manchester University.

―― (1969b). *J. Appl. Phys.*, **40**, 4675.

―― (1973). *J. Phys. D: Appl. Phys.*, **6**, 1358.

―― and Rhoderick, E.H. (1969). *J. Phys. D: Appl. Phys.*, **2**, 465.

―― ―― (1971). *Solid-St. Electron.*, **14**, 71.

Spenke, E. (1958). *Electronic semiconductors*. McGraw-Hill, New York.

Spitzer, W.G. and Mead, C.A. (1963). *J. Appl. Phys.*, **34**, 3061.

Stokoe, T.Y. and Parrott, J.E. (1974). *Solid-St. Electron.*, **17**, 477.

Stratton, R. (1962). *Phys. Rev.*, **126**, 2002.

Strikha, V.I. (1964). *Radio-Engng. Electron. Phys.* (USSR) **4**, 552.

Sullivan, A.B.J. (1976). *Electron. Lett.*, **12**, 133.

Sullivan, M.W. and Eigler, J.H. (1957). *J. Electrochem. Soc.*, **104**, 226.

Sze, S.M. (1969). *Physics of semiconductor devices*. John Wiley & Sons, New York.

―― Coleman, D.J., and Loya, A. (1971). *Solid-St. Electron.*, **14**, 1209.

―― Crowell, C.R., and Kahng, D. (1964). *J. Appl. Phys.*, **36**, 2534.

―― and Gibbons, G. (1966). *Solid-St. Electron.*, **9**, 831.

Szydlo, N. and Poirier, R. (1973). *J. Appl. Phys.*, **44**, 1386.

Tejedor, C., Flores, F., and Louis, E. (1977). *J. Phys. C. (Solid State Phys.)*, **10**, 2163.

Thanailakis, A. (1974). *Proc. Manchester Conf. Metal—Semicond. Contacts*, p. 59. Inst. Phys., London.

—— (1975). *J. Phys. C. (Solid State Phys.)*, **8**, 655.

—— and Northrop, D.C. (1971). *J. Phys. D: Appl. Phys.*, **4**, 1776.

—— —— (1973). *Solid-St. Electron.*, **16**, 1383.

—— and Rasul, A. (1976). *J. Phys. C. (Solid State Phys.)*, **9**, 337.

Todd, C.J., Ashwell, G.W.B., Speight, J.D., and Heckingbottom, R. (1974). *Proc. Manchester Conf. Metal—Semicond. Contacts*. Inst. Phys., London.

Torrey, H.C. and Whitmer, C.A. (1948). *Crystal rectifiers*. McGraw-Hill, New York.

Tove, P.A., Hyder, S.A., and Susila, G. (1973). *Solid-St. Electron.*, **16**, 513.

Tu, K.N., Alessandrini, E.J., Chu, W.K., Kraütle, H., and Mayer, J.W. (1974). *Jap. J. Appl. Phys. Suppl.*, **2-1**, 669.

Turner, M.J. and Rhoderick, E.H. (1968). *Solid-St. Electron.*, **11**, 291.

Van Laar, J. and Sheer, J.J. (1967). *Surf. Sci.*, **8**, 342.

Viktorovich, P. and Kamarinos, G. (1976). *Solid-St. Electron.*, **19**, 1041.

Vilms, J. and Wandinger, L. (1969). *Ohmic contacts to semiconductors*, p. 31. Electrochemical Society, New York.

Vincent, G., Bois, D., and Pinard, P. (1975). *J. Appl. Phys.*, **46**, 5173.

Wagner, C. (1931). *Phys. Z.*, **32**, 641.

Walker, L.G. (1974). *Solid-St. Electron.*, **17**, 763.

Walpole, J.N. and Nill, K.W. (1971). *J. Appl. Phys.*, **42**, 5609.

Wilkinson, J.M. (1974). *Proc. Manchester Conf. Metal—Semicond. Contacts*, p. 27. Inst. Phys., London.

—— Wilcock, J.D., and Brinson, M.E. (1977). *Solid-St. Electron.*, **20**, 45.

Williams, R. (1966). *J. Appl. Phys.*, **37**, 3411.

Wilson, A.H. (1932). *Proc. Roy. Soc. A*, **136**, 487.

Wronski, C.R., Carlson, D.E., and Daniel, R.E. (1976). *Appl. Phys. Lett.*, **29**, 602.

Yndurain, F. (1971). *J. Phys. C. (Solid State Phys.)*, **4**, 2849.

—— and Rubio, J. (1971). *Phys. Rev. Lett.*, **26**, 138.

Yu, A.Y.C. (1970). *Solid-St. Electron.*, **13**, 239.

—— and Snow, E.H. (1968). *J. Appl. Phys.*, **39**, 3008.

—— —— (1969). *Solid-St. Electron.*, **12**, 155.

Zettlemoyer, A.C. (1969). *Ohmic contacts to semiconductors*, p. 48. Electrochemical Society, New York.

Ziegler, J.F., Mayer, J.W., Kircher, C.J., and Tu, K.N. (1973). *J. Appl. Phys.*, **44**, 3851.

Zohta, Y. (1970). *Appl. Phys. Lett.*, **17**, 284.

—— (1973). *Solid-St. Electron.*, **16**, 1029.

Index